Mobile Test Automation with Appium

Comprehensive guide to build mobile test automation solution using Appium

Nishant Verma

BIRMINGHAM - MUMBAI

Mobile Test Automation with Appium

First published: June 2017

Production reference: 1290617

Published by Packt Publishing Ltd.
Livery Place
35 Livery Street
Birmingham
B3 2PB, UK.
ISBN 978-1-78728-016-8

www.packtpub.com

Credits

Author
Nishant Verma

Reviewers
Jagannath Balachandran
Kapil Sethi
Manoj Hans

Commissioning Editor
Amarabha Banerjee

Acquisition Editor
Siddharth Mandal

Content Development Editor
Aditi Gour

Technical Editor
Rashil Shah

Copy Editor
Shaila Kusanale

Project Coordinator
Ritika Manoj

Proofreader
Safis Editing

Indexer
Tejal Daruwale Soni

Graphics
Jason Monteiro

Production Coordinator
Arvindkumar Gupta

About the Author

Nishant Verma is a co-founder of TestVagrant Technologies. It's a service start-up addressing testing solutions for B2C companies with a focus on mobile and web, and helps companies deliver faster and reliably.

Nishant has 11 years of experience in software development and testing. He has worked with IT companies such as ThoughtWorks Inc., Aditi Technologies, and Altisource. He has extensive experience in setting up agile testing practices, functional and non-functional test automation, mentoring, and coaching. In the past, he has worked on web UIs and specializes in building test solutions in the mobile domain. He has hands-on experience with test automation tools such as WebDriver (Selenium2), Calabash, Frank, Appium, Watin, Sikuli, QTP, and DeviceAnywhere.

He actively maintains his own website on testing techniques, agile testing, automation techniques, and general learning. He has contributed to leading testing journals such as Testing Circus and Software Developer's Journal, and has been an active speaker at vodQA (testing event of Thoughtworks).

Nishant has authored a reference book on how to use Appium for automating Android apps using Java, which is available on Gitbook. It has received close to 200,000 views, 40,000 readers online, and has been downloaded around 3,000 times.

About the Reviewers

Jagannath Balachandran works as a lead consultant for ThoughtWorks India Pvt. Ltd. He has around 14 years of experience working with teams delivering software using agile and continuous delivery practices. He has extensively consulted clients on their journey toward continuous delivery.

Kapil Sethi is an agile practitioner with more than 12 years of experience in the software industry. He is a passionate advocate of shifting testing to the left most column in the agile development process and is a strong believer of the Testing Pyramid. He is a connoisseur of automation testing and has hands-on experience in designing automation testing frameworks using a variety of automation tools, such as WebDriver, Appium, Protractor, Applitools, Calabash, SoapUI, and QTP.

He has worked on numerous domains, including banking, mortgage, retail, e-commerce, and online gaming. His expertise involves helping development teams deliver quality products, coaching teams on agile adoption, transforming teams and thereby organizations, to make the working environment fun and passionate.

He is currently working with Nintex as an automation specialist. In the past, he has worked with companies such as MYOB, ThoughtWorks, Sapient, and Cognizant Technology Solutions, and performed the development lead, iteration manager, QA lead roles during his tenure.

www.PacktPub.com

For support files and downloads related to your book, please visit www.PacktPub.com.

Did you know that Packt offers eBook versions of every book published, with PDF and ePub files available? You can upgrade to the eBook version at www.PacktPub.com and as a print book customer, you are entitled to a discount on the eBook copy. Get in touch with us at service@packtpub.com for more details.

At www.PacktPub.com, you can also read a collection of free technical articles, sign up for a range of free newsletters and receive exclusive discounts and offers on Packt books and eBooks.

https://www.packtpub.com/mapt

Get the most in-demand software skills with Mapt. Mapt gives you full access to all Packt books and video courses, as well as industry-leading tools to help you plan your personal development and advance your career.

Why subscribe?

- Fully searchable across every book published by Packt
- Copy and paste, print, and bookmark content
- On demand and accessible via a web browser

Customer Feedback

Thanks for purchasing this Packt book. At Packt, quality is at the heart of our editorial process. To help us improve, please leave us an honest review on this book's Amazon page at `https://www.amazon.com/dp/1787280160`.

If you'd like to join our team of regular reviewers, you can e-mail us at `customerreviews@packtpub.com`. We award our regular reviewers with free eBooks and videos in exchange for their valuable feedback. Help us be relentless in improving our products!

Table of Contents

Preface

With the growing popularity of mobile apps and the enormous growth in the number of mobile devices all around the world, mobile ecosystems are poised to further scale up. Until a couple of years ago, the IT world was dominated by web and enterprise application development and testing. With the growth of mobile apps around the world, the trend is shifting toward mobile development and testing as a niche skill set. Mobile testing had largely been manual until the advent of standard test automation libraries, such as Calabash and Appium.

This book is an effort toward gearing up a better testing workforce by making them educated and aware of a mobile testing and automation tool called Appium. Appium is the most widely adopted mobile test automation tool. The community support has been vibrant, but there is a lack of a structured step-by-step guide or documentation around building a framework. This book is an attempt to bridge that gap and serves as a handheld guide for each tester who wants to build their own mobile test automation framework from scratch.

This book is intended for developers and testers who want to learn mobile app testing and automation using Appium. The book takes you on a journey of understanding Appium and slowly gets you started with the test automation ecosystem. Cucumber is one of the most promising technologies, and is rising in popularity due to the wide adoption of the agile and behavior-driven development methodologies. This book introduces you to the concept of Cucumber and shows how you can build your own testing framework in Cucumber and Appium from scratch. It contains example code snippets of creating a sample project, writing first Appium tests, evolving the test framework, and setting up Jenkins.

The book is organized into two parts:

- **Appium basics**: This largely covers an understanding of Appium, desired capabilities in Appium, Appium inspector, and how to use it to find locators, test synchronization, and automate widely used gestures, such as tap, scroll, press, and long press.
- **Appium advanced**: This covers design patterns for the automation framework, how to run tests on actual devices and emulators, how to run tests on a Genymotion emulator, continuous integration with Jenkins, and Appium tips and tricks.

What this book covers

Chapter 1, *Introduction to Appium*, starts with an introduction to the mobile app. It talks about different types of mobile app, that is, native, hybrid, and mobile web. We then take a little closer look at the advantages and limitations of each type of mobile app. We learn about Appium's architecture and about two different automation frameworks Appium uses, XCUITest and UIAutomator2 for iOS and Android, respectively.

Chapter 2, *Machine Setup*, starts with instructions for setting up your machine in order to start using Appium and write automated tests. In this chapter, we address the setup for both Windows and Mac machines. Some of the prerequisites to install are the most recent Java, Android SDK, Genymotion Emulator, Appium, IntelliJ as the preferred IDE, and the app under test. We will also learn to create the sample Android emulator as well as the Genymotion emulator. We will learn how to install Appium, both via npm and the Appium GUI app. We will take a detailed look at the Appium GUI app and the iOS and Android settings Appium allows.

Chapter 3, *Writing Your First Appium Test*, helps us write our first Appium test. We will start by creating a Java project in IntelliJ and then get introduced to Cucumber. We create a sample feature file and write our first scenario using the Given-When-Then format. We will learn how to start Appium session and use Appium Inspector. We will then write our first automated test and learn how to run the cucumber test. We will also learn how to write our first test for mobile web app and learn how to use the Chrome developer tools to find the locators. We then run these tests via the IDE.

Chapter 4, *Understanding Desired Capabilities*, tells us about the concept of desired capabilities in Appium. We learn about the mandatory capabilities and the device-specific desired capabilities, such as Android and iOS. We will look into the server argument and the various flags it exposes along with its sample usage.

Chapter 5, *Understanding Appium Inspector to Find Locators*, shows us how to use the Appium inspector to find the locator of a UI element. We learn to derive the xPath over the Appium-generated xPath values. We looked into another tool, UIAutomatorViewer, and how to use it. We also learn how to debug the mobile apps using Chrome's inspect feature.

Chapter 6, *How to Synchronize Tests*, explores the different types of drivers Appium allows you to create, along with the various synchronization strategies. We will learn about the implicit wait, explicit wait, and fluent wait. We also learn about ExpectedConditions and the various predefined conditions it allows.

`Chapter 7`, *How to Automate Gestures*, explains implementing various gestures that Appium supports. We will learn how to implement the most frequently used gestures, such as tap, swipe, scroll, and drag and drop. We will also learn about the orientation and how to change the orientation of devices between the landscape and portrait modes.

`Chapter 8`, *Design Patterns in Test Automation*, covers the concept of the design pattern in test automation. In this chapter, we will take a detailed look at the page object pattern and then learn how to implement it in the current framework, which we have been building since `Chapter 3`, *Writing Your First Appium Test*. We will learn about assertions and where they belong, and we will also learn about the concepts of setup and teardown and how to implement them using pre-specified hooks in cucumber.

`Chapter 9`, *How to Run Appium Test on Devices and Emulators*, shows you how to connect physical devices and prepare them for development and testing purposes. It also demonstrates how to configure the Genymotion emulator and run tests. We learn how to retrieve the UDID of iOS devices, the libraries we need to install, and the process for running the test on iOS.

`Chapter 10`, *Continuous Integration with Jenkins*, teaches the concept of Gradle and writing Gradle tasks. We start by creating a Gradle task to run the test via command line and moving the project to Git. We navigate through downloading and installing Jenkins. We learn how to set up a Jenkins job and trigger it and view the report. This chapter explains how to implement continuous testing using Jenkins.

`Chapter 11`, *Appium Tips and Tricks*, shows you some tips and tricks in the form of code snippets, that can be used to make your test framework more intelligent and innovative. We will learn about switching between webviews and native views, taking screenshots, and recording video using adb commands. We will also explore the approach of running tests in parallel on multiple devices and about the network simulation API.

`Chapter 12`, *Appium Desktop App*, explores the new Appium Desktop App. It discusses in details about how to install the new Appium GUI app, how to start an appium server with basic and advanced options. It also explains how to use the Desired Capabilities while setting up a session and how to connect to different end points (the non local server).

By the end of this book, you will have learned about Appium, how to build a test automation framework from scratch in Cucumber and Appium, and how to set up Jenkins to run tests.

`Appendix`, Appendix takes a deeper insight into different how to's which are needed across chapters. It includes a deep dive into Cucumber and explains various concepts of Cucumber. It also talks about finding details needed for Appium for android installer. Very importantly, it tells us how to install the Google Play services on the Genymotion Emulator.

What you need for this book

To get started with this book, you need basic knowledge of Java. You should be aware of the OOPS concept and should be able to use loops and define classes. A basic understanding of mobile apps and knowledge of Android would be an added advantage; however, it is not a must. The book provides hands-on experience with writing and executing code. There are some software prerequisites, which are explained in the second chapter, which helps set up the development environment and readies your machine for any future mobile automation work using Appium.

Who this book is for

This book is intended for developers/testers who want to learn mobile automation using Appium. It doesn't require any prior experience in testing mobile applications or automation. This book serves as a detailed guide for Appium and a step-by-step guide to building a mobile test automation framework from scratch. The only prerequisite for this book is to have a basic knowledge of Java programming. By the end of this book, you would have gained advanced knowledge of Appium and would have learned how to build a framework in Cucumber and Appium. You will be able to leverage this framework building knowledge by replacing Appium with any other UI automation tool, such as Selenium.

Conventions

In this book, you will find a number of text styles that distinguish between different kinds of information. Here are some examples of these styles and an explanation of their meaning.

Code words in text, database table names, folder names, filenames, file extensions, pathnames, dummy URLs, user input, and Twitter handles are shown as follows: "With version 1.6, Appium has provided support to UiAutomator 2. Appium uses the `appium-android-bootstrap` module to interact with UI Automator. "

A block of code is set as follows:

```
@Before
public void startAppiumServer() throws IOException {
    AppiumDriverLocalService appiumService =
    AppiumDriverLocalService.buildDefaultService();
    appiumService.start();
}
```

Any command-line input or output is written as follows:

```
automationName: XCUITest
```

New terms and **important words** are shown in bold. Words that you see on the screen, for example, in menus or dialog boxes, appear in the text like this: "Clicking on **Start Session** will launch a new Appium inspector screen, as illustrated."

Warnings or important notes appear in a box like this.

Tips and tricks appear like this.

Reader feedback

Feedback from our readers is always welcome. Let us know what you think about this book-what you liked or disliked. Reader feedback is important for us as it helps us develop titles that you will really get the most out of.

To send us general feedback, simply e-mail `feedback@packtpub.com`, and mention the book's title in the subject of your message.

If there is a topic that you have expertise in and you are interested in either writing or contributing to a book, see our author guide at `www.packtpub.com/authors`.

Customer support

Now that you are the proud owner of a Packt book, we have a number of things to help you to get the most from your purchase.

Downloading the example code

You can download the example code files for this book from your account at `http://www.p acktpub.com`. If you purchased this book elsewhere, you can visit `http://www.packtpub.c om/support` and register to have the files e-mailed directly to you.

You can download the code files by following these steps:

1. Log in or register to our website using your e-mail address and password.
2. Hover the mouse pointer on the **SUPPORT** tab at the top.
3. Click on **Code Downloads & Errata**.
4. Enter the name of the book in the **Search** box.
5. Select the book for which you're looking to download the code files.
6. Choose from the drop-down menu where you purchased this book from.
7. Click on **Code Download**.

Once the file is downloaded, please make sure that you unzip or extract the folder using the latest version of:

- WinRAR / 7-Zip for Windows
- Zipeg / iZip / UnRarX for Mac
- 7-Zip / PeaZip for Linux

The code bundle for the book is also hosted on GitHub at `https://github.com/PacktPubl ishing/Mobile-Test-Automation-with-Appium`. We also have other code bundles from our rich catalog of books and videos available at `https://github.com/PacktPublishing/`. Check them out!

Errata

Although we have taken every care to ensure the accuracy of our content, mistakes do happen. If you find a mistake in one of our books-maybe a mistake in the text or the code-we would be grateful if you could report this to us. By doing so, you can save other readers from frustration and help us improve subsequent versions of this book. If you find any errata, please report them by visiting `http://www.packtpub.com/submit-errata`, selecting your book, clicking on the **Errata Submission Form** link, and entering the details of your errata. Once your errata are verified, your submission will be accepted and the errata will be uploaded to our website or added to any list of existing errata under the Errata section of that title.

To view the previously submitted errata, go to https://www.packtpub.com/books/content/support and enter the name of the book in the search field. The required information will appear under the **Errata** section.

Piracy

Piracy of copyrighted material on the Internet is an ongoing problem across all media. At Packt, we take the protection of our copyright and licenses very seriously. If you come across any illegal copies of our works in any form on the Internet, please provide us with the location address or website name immediately so that we can pursue a remedy.

Please contact us at copyright@packtpub.com with a link to the suspected pirated material.

We appreciate your help in protecting our authors and our ability to bring you valuable content.

Questions

If you have a problem with any aspect of this book, you can contact us at questions@packtpub.com, and we will do our best to address the problem.

1
Introduction to Appium

The mobile app market is huge, and it will increase further. Approximately, there are 2 billion smartphone devices worldwide, which is more than two times the number of personal computers in the world. A report (for more information, visit `https://www.statista.com/topics/1002/mobile-app-usage/`) shows that more than 102 billion apps have been downloaded worldwide, and the number is expected to reach 268 billion by 2017. According to one of the reports (for more information, visit `http://www.statista.com/statistics/269025/worldwide-mobile-app-revenue-forecast/`), the worldwide mobile revenue for 2015 amounted to $41.1 billion and is expected to reach $101.1 billion by 2020.

With all these promising growth numbers and trends, learning mobile app development and testing will be worth it and will have a huge demand.

In this chapter, we will cover the following topics:

- Types of mobile apps
    - Native App
    - Mobile Web app
    - Hybrid App
- Appium Architecture
    - What is XCUITest
    - What is UiAutomator 2

Let's take a look at mobile apps, which form this ecosystem, and how they are broadly categorized based on the way they are developed:

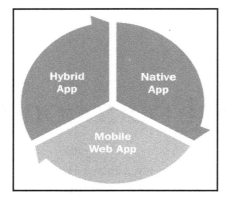

Let's understand the different types of mobile apps.

The mobile world is majorly dominated by two operating systems: iOS and Android. Most apps are made for both the platforms given the user base. In this chapter, we will take a detailed look at the following:

- Native app, mobile web, and hybrid app
- The characteristics of each type of app
- A sample example app of each type

Native app

A native app is an app developed for a particular mobile device or platform (such as Android, iOS, or Windows). For example, iPhone apps are written in Swift, and Android apps are written in Java. Native apps are also better performing and have a high degree of reliability as they use the underlying system architecture and the device's built-in features.

Native apps can run in both the online mode as well as the offline mode. Native App is tied to the mobile operating system it has been developed for, and hence can't be run on any other operating system. This makes developing the native app costly as the same app has to be rewritten for another operating system. These apps are available to be downloaded on the mobile via the respective app store.

Here's an example of a native app. It's a news app bundled with iPhone and can be downloaded from the Apple App Store:

Another one is the popular Instagram app on Android phone, which is native:

Mobile Web app

A Mobile Web app is an app accessed over a mobile browser. It can be easily accessed via built-in browsers, such as Safari on iOS and Chrome on Android. They are primarily developed using technologies such as HTML5 or JavaScript, which provide the customization capabilities. So, they are basically served from a server and not stored offline anywhere on the device.

Web apps have a common code base and can be accessed like any typical web app on any device with browsers. For Mobile Web apps, responsive web design is the new standard as they have to cater to devices of different screen sizes and resolutions. Mobile Web apps can also access mobile-specific features, such as dialing a phone number or location-based mapping. Mobile Web apps can only be accessed with a valid network (Wifi/4G/3G/2G).

The following is an example of a mobile app. It's a mobile website of The New York Times and can be opened with any mobile browser. The URL for this is `http://mobile.nytimes.com`. One can perform the same actions as web, such as browser refresh. The following screenshot shows the same app; it's opened using the Safari app on an iPhone 6 simulator, running iOS 10.1:

The next is an Android emulator running Android 6.0 and has the mobile site of The New York Times opened on the default browser app:

Hybrid app

A hybrid app consists basically of websites packaged in a native wrapper. They are primarily developed in web technologies (HTML5, CSS, JavaScript) but run inside a native container, thereby giving a feel that it is a native app. Hybrid apps rely on HTML being rendered in the mobile browser, with a limitation that the browser is embedded within the app. This approach allows you to have one code base for all the mobile operating systems: iOS, Android, and Windows. A web-to-native abstraction layer enables access to device-specific capabilities, which are not accessible in Mobile Web apps otherwise. Examples include a camera, on device local storage, and an accelerometer.

Hybrid app is the most favored approach for companies with a web page in existence. Those companies often build hybrid apps as a wrapper over the web page. Tools such as PhoneGap and Sencha Touch allow one to build a hybrid app. These apps can be downloaded via the respective app stores. Here's an example of a hybrid app--it's an Evernote app and can be downloaded from the respective app store:

The mobile testing ecosystem is not yet crowded; there are only a couple of tools that are really worth trying and learning, and Appium is the most promising one.

Appium is an open source tool to automate mobile applications. It's a cross-platform automation tool, which will help in automating the different types of mobile apps that we discussed earlier.

The supported mobile operating system platforms by Appium are as follows:

* iOS
* Android
* Windows

Let's take a detailed look at Appium, how it is architected, and how it facilitates automation.

Appium architecture

Now that we have understood the different types of mobile apps, let's take a look at how Appium is architected to support mobile app automation. Appium is basically a web server written in Node.js. The server performs actions in the given order:

* Receives connection from client and initiates a session
* Listens for commands issued
* Executes those commands
* Returns the command execution status

So basically, Appium is a client-server architecture.

The Appium server receives a connection from client in the form of a JSON object over HTTP. Once the server receives the details, it creates a session, as specified in JSON, and returns the session ID, which will be maintained until the Appium server is running. So, all testing will be performed in the context of this newly created session. The following is a diagram depicting the Appium architecture:

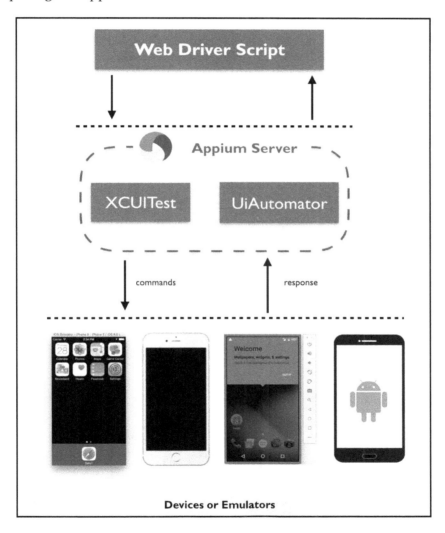

The Appium server is written in Node.js; it can be installed via npm or directly from source.

XCUITest

XCUITest is an automation framework introduced by Apple with the iOS 9.3 version. However, from iOS 10 and later versions, it's the only supported automation framework.

Appium 1.6.0 uses Apple's new XCUITest framework, which supports iOS 10/Xcode 8. Appium internally uses Facebook's WebDriverAgent project to support XCUITest. For the older iOS version (<=9.3), Appium uses Apple's UIAutomation library. Typical usage would be to pass the following in desired capabilities:

```
automationName: XCUITest
```

Facebook WebDriverAgent is a WebDriver server implementation for iOS. It is used to remote control connected devices or simulators and allows one to launch an app, perform commands (such as tap and scroll), and kill applications.

The UIAutomation library communicates with bootstrap.js, which is running inside the device or simulator to perform the commands received by the Appium client:

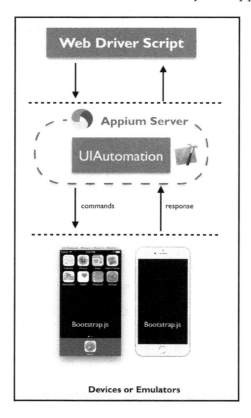

UiAutomator 2

UIAutomator 2 is an automation framework based on Android instrumentation and allows one to build and run UI tests.

Appium uses Google's UIAutomator to execute commands on real devices and emulators. UIAutomator is Google's test framework for native app automation at the UI level. Typical usage would be to pass the following in desired capabilities:

```
automationName: uiautomator2
```

With version 1.6, Appium has provided support to UiAutomator 2. Appium uses the `appium-android-bootstrap` module to interact with UI Automator. It allows commands to be sent to the device, which are then executed on real devices using Android's UIAutomator testing framework.

When Appium client requests to create a new `AndroidDriver` session, the client passes the desired capability to the Appium node server. The UIAutomator2 driver module creates the session. It then installs the UIAutomator2 server APK on the connected Android device, starts the Netty server, and initiates a session. Once the Netty server session is started, the UIAutomator2 server continues to listen on the device for requests and responds:

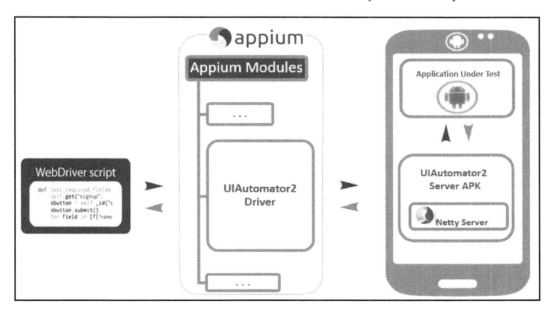

Picture courtesy--GitHub Appium page

Pros of using Appium

Appium has many advantages; some of them are listed here:

1. It's an open source tool backed by a very active community.
2. It supports multiple languages (Java, JavaScript, Objective C, C#, PHP, Python, Ruby, Clojure, and Perl).
3. It doesn't force you to recompile an app or modify it. You can test the same version that you have planned to submit to the play store or app store.
4. It allows you to write cross-platform tests.

 Netty is an NIO client-server framework, which enables quick and easy development of network applications, such as protocol servers and clients.

Summary

So, in this chapter, we learned about different types of mobile apps and the advantages one has over another. We also learned about Appium and its architecture. We learned about XCUITest and UIAutomator 2 and how Appium uses them to communicate commands to devices. We also looked at the advantages of using Appium.

In the next chapter, we will set up our machine so that we can start practicing the usage of Appium.

2
Setting Up the Machine

In the last chapter, we looked at the different types of mobile apps. We also looked at the advantage of one over another and how they are different from each other. We learned about Appium and its architecture, and we learned about iOS XCUITest and android UIAutomator 2.

We also learned how commands are translated and passed on to the device. In the upcoming chapters, we will learn how to set up the machine and start writing tests and how to eventually create a framework.

In this chapter, we will cover the following topics:

- Installing Java
- Installing Android SDK and creating one Android Virtual Device
- Installing Genymotion Emulator
- Installing Appium (Via NPM, app, source code)
- Choosing IDE and setting up
- Knowing app under test

All the preceding installations are mandatory, except some that are optional and indicated. As part of this book, we will be addressing both Mac and Windows machines.

Machine setup for macOS

Setting up the machine will require a bunch of software and packages to be installed. Let's start with `bash_profile`. Open the terminal and type in the following command (in the home directory):

```
ls -al
```

This should return all the hidden files and directories under the home directory. Check whether the `.bash_profile` file is present; if not, type the given command to create one:

```
touch .bash_profile
```

Installing Java

If you have had the development machine set up before, you might have a couple of software and packages already installed. You can skip the installation part and check for the version of the installed packages. If the versions are significantly old, you might want to upgrade them.

For the new machines, follow the mentioned steps for installing Java:

1. Visit the JDK download page and download the `jdk-8uversion-macosx-xxx.dmg` package based on your machine configuration (either the amd64 or x64).
2. Install Java from the downloaded package.
3. Once installed, launch the terminal and type in this command to determine the Java version:

    ```
    java -version
    ```

4. You will see the following output if Java is installed correctly:

    ```
    java version "1.8.0_73"
    Java(TM) SE Runtime Environment (build 1.8.0_73-b02)
    Java HotSpot(TM) 64-Bit Server VM (build 25.73-b02, mixed mode)
    ```

5. Add the following line to your `.bash_profile` file. In Mac OSX 10.5 or later versions, Apple recommends to set the `$JAVA_HOME` variable to `/usr/libexec/java_home`:

```
export JAVA_HOME=$(/usr/libexec/java_home)
export PATH=$PATH:$JAVA_HOME/bin
```

Now that we have finished installing Java, let's move on to installing Android SDK.

Installing Android SDK (using the Android command-line tool)

1. Navigate to the Android Studio page and download the command-line tools (for more information visit this link: `https://developer.android.com/studio/index.html?hl=sk`).
2. Once downloaded, extract the same in a folder of your choice.
3. Rename the extracted file, for your convenience, to Android SDK.
4. Android SDK contains only the basic SDK tools and does not contain any platform or library; we need to download the same before we start using it:
   1. Launch the terminal and navigate to the folder where the ZIP file was extracted. In the terminal, type `android` and press enter.
   2. Android SDK Manager will start with a new window.
   3. Select one of the android platform **Android 7.0 (API 24)** and choose the given packages: **ARM EABI v7a System Image**, **Intel x86 Atom System Image**, and **SDK Platform**.
   4. Under the **Tools** section, select **Android SDK**, **Android SDK Platform-tools**, and **Android SDK Build-tools**.
   5. Under the **Extra** section, select **Google Play Services**.

5. Open the `.bash_profile` file and enter the following lines at the bottom:

```
export ANDROID_HOME={YOUR_PATH}
export PATH=$PATH:$ANDROID_HOME/tools:
$ANDROID_HOME/platform-tools
```

6. Save the file and run this command:

```
source ~/.bash_profile
```

7. Run the next command to check whether the Android home is set properly:

```
echo $ANDROID_HOME
```

Installing Android SDK (using Homebrew) (Optional)

You can also choose to install Android SDK using Homebrew (the `brew install android-sdk` command). This installs the Android SDK in the `/usr/local/Cellar/android-sdk/{YOUR_SDK_VERSION_NUMBER}` path, so `ANDROID_HOME` should point to the installed location.

Creating Android Virtual Device (Optional)

When we install Android SDK, it allows you to create a virtual device (an emulator) to help perform the development and testing locally without buying a physical device. The following steps will help you to create an emulator:

1. Launch AVD Manager (using the terminal, type in command `android avd`).
2. Click on "**Create...**".
3. Enter an **AVD Name**.
4. Select a target **Device** from the dropdown.
5. Select an **API level** by clicking on the dropdown next to **Target**.
6. Click on the dropdown next to **CPU/ABI** and select a value from the dropdown.

7. Choose a **Skin**.
8. You can alter the **RAM** size in **Memory Options**; it generally defaults based on the device selected.
9. Press **OK**:

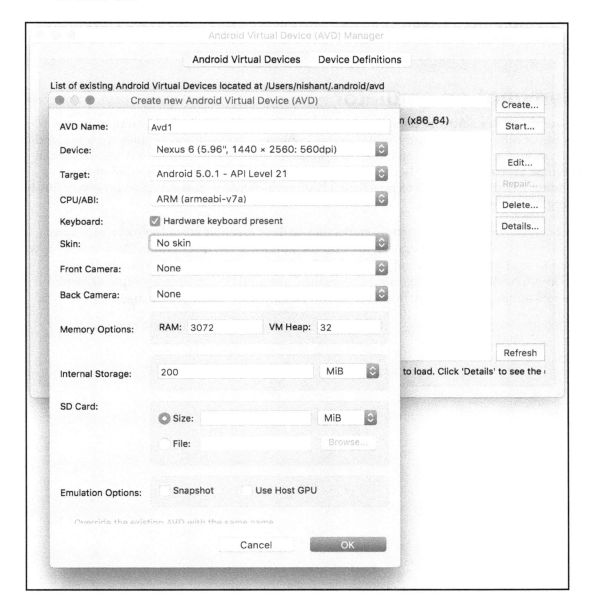

Once the Android Virtual Device is created, it will show up in the AVD manager. We have the option of performing operations, such as Start, Edit, Repair, and Delete, on the Android Virtual Device.

Let's take a look at an emulator (Genymotion) other than Android Virtual Device, which is a much better performant compared to the android ones. We can rely heavily on the Genymotion emulator for our day-to-day testing and development activities.

Genymotion emulator

Genymotion is a software company making one of the fastest Android emulators and a couple of other products around it. For this book and our testing activities, we will use the Genymotion emulator personal use version, as it's a faster alternative to Android Virtual Devices. However, I strongly recommend that you get in touch with Genymotion to validate whether you need to purchase a license (individual or enterprise) once you finish using the personal version.

One can download the Genymotion installer to their machine from the website `https://www.genymotion.com/` after signing up, and then perform the installation. Once installed, we need to sign in with Genymotion account details. Post that, we need to create virtual devices. The illustrated screenshot shows the Genymotion app with a couple of virtual devices already configured; however, it will be empty for the first time user:

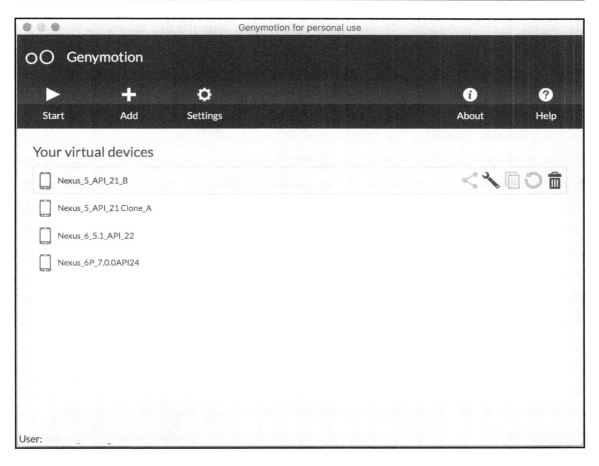

Let's learn to create virtual devices in Genymotion:

1. Launch the Genymotion app and log in with your registered credentials.
2. Click on **Add** +.

3. Select the **Android version** and **Device model** from the dropdown:

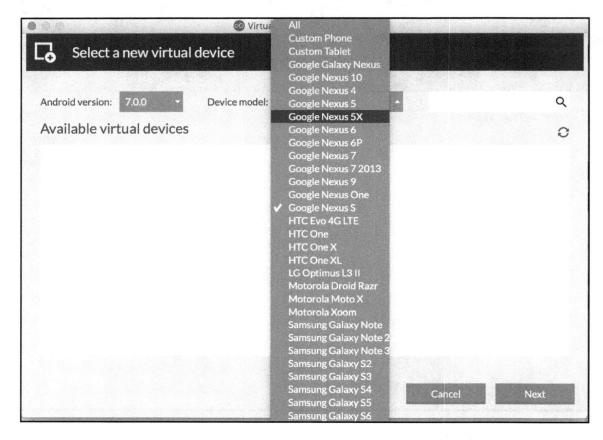

4. Select **Android version** as **5.1** and choose **Device Model** to be **Google Nexus 6**.
5. Click on **Next**.
6. It will start the download of the virtual device.
7. Once done, click on **Finish** to close the download window.
8. The virtual device will start appearing on the **Your virtual devices** window.
9. Click on **Start** to launch the virtual device.

Starting the emulator will show a screen similar to the following:

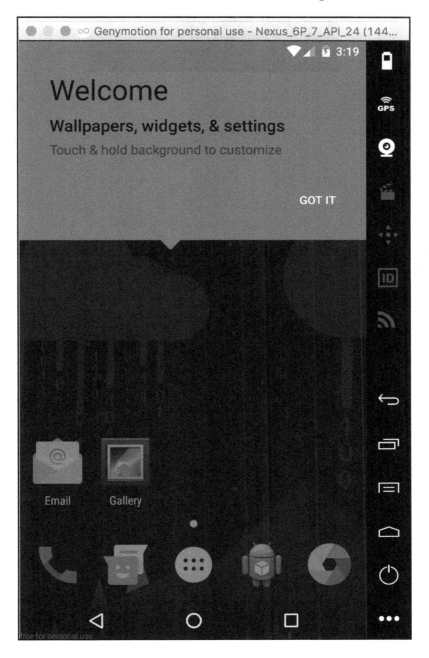

Launch the terminal and run the `adb devices` command; the output will be as shown:

```
[→ ~ adb devices
List of devices attached
192.168.56.101:5555      device
```

So, each emulator runs as a virtual machine on your physical machine. To install the app on Genymotion emulator, the normal adb commands--`adb install /path/to/app/<app_name>.apk` will work fine.

Debug help

If the `adb devices` command throws the error `adb server version (31) doesn't match this client (36)`, follow the given set of steps to fix the same:

1. Navigate to **Genymotion** > **Settings** > **ADB** > **Use custom Android SDK tools**.
2. Put the `ANDROID SDK` path in **Android SDK**:

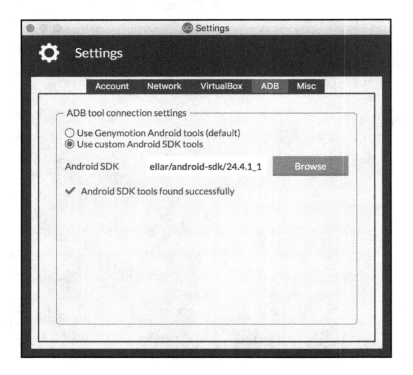

3. Once done, restart the Genymotion virtual device and type in the `adb devices` command.

This should help fix it.

Installing Appium

Appium requires macOS X 10.7 or a higher version; we would recommend 10.10 (Yosemite) or a later version. To work with Appium, we need both the Appium GUI app (installed via the `.dmg` file) and the Appium server (installed via Homebrew), as explained here:

1. Install Xcode and the Xcode command-line tool.
2. Download the `appium.dmg` file from `https://bitbucket.org/appium/appium.app/downloads/`.
3. Install Homebrew (`http://brew.sh/`) on your machine. This will need Ruby to be installed on your machine. Run the below command to install Homebrew:

 Homebrew is a package manager for Mac OS. It installs packages to their own directories and then symlinks their files into `/usr/local`.

```
/usr/bin/ruby -e "$(curl -fsSL
https://raw.githubusercontent.com/Homebrew/install/master/install)"
```

4. Post the install run command to ensure that Homebrew is up to date:

```
brew update
```

5. To install node on your machine, run this command:

```
brew install node
```

6. Once the preceding commands are executed and don't throw up any errors, we can run one final command to check the successful Node and npm install. Run the below command to check the successful installation (of Node and npm) and this should successfully install `grunt-cli`:

Node.js is an open source, cross-platform JavaScript runtime environment for developing a diverse variety of tools and applications.
npm is the default package manager for the JavaScript runtime environment Node.js.

7. To install Appium server and Appium doctor using node, we can run the given command:

```
npm install -g appium
npm install -g appium-doctor
```

8. Once Appium doctor is installed, we can run the `appium-doctor` command in the terminal and see the following output:

```
→ ~ appium-doctor
info AppiumDoctor Appium Doctor v.1.2.5
info AppiumDoctor ### Diagnostic starting ###
info AppiumDoctor   ✔ Xcode is installed at: /Applications/Xcode.app/Contents/Developer
info AppiumDoctor   ✔ Xcode Command Line Tools are installed.
info AppiumDoctor   ✔ DevToolsSecurity is enabled.
info AppiumDoctor   ✔ The Authorization DB is set up properly.
info AppiumDoctor   ✔ The Node.js binary was found at: /usr/local/bin/node
info AppiumDoctor   ✔ Carthage was found at: /usr/local/bin/carthage
info AppiumDoctor   ✔ HOME is set to: /Users/nishant
info AppiumDoctor   ✔ ANDROID_HOME is set to: /usr/local/Cellar/android-sdk/24.4.1_1
info AppiumDoctor   ✔ JAVA_HOME is set to: /Library/Java/JavaVirtualMachines/jdk1.8.0_73.jdk/Contents/Home
info AppiumDoctor   ✔ adb exists at: /usr/local/Cellar/android-sdk/24.4.1_1/platform-tools/adb
info AppiumDoctor   ✔ android exists at: /usr/local/Cellar/android-sdk/24.4.1_1/tools/android
info AppiumDoctor   ✔ emulator exists at: /usr/local/Cellar/android-sdk/24.4.1_1/tools/emulator
info AppiumDoctor   ✔ Bin directory of $JAVA_HOME is set
info AppiumDoctor ### Diagnostic completed, no fix needed. ###
info AppiumDoctor
info AppiumDoctor Everything looks good, bye!
info AppiumDoctor
→ ~
```

9. Once done, we can run the `appium` command in the terminal and see this output:

```
→ ~ appium
[Appium] Welcome to Appium v1.6.0-beta3 (REV d345483de164e062727f95d3f
bd575433d735548)
[Appium] Appium REST http interface listener started on 0.0.0.0:4723
```

So, the preceding section completes the Appium setup on both the app and the server. However, we can also install Appium server from source.

Installing Appium server (From Source) (Optional)

This is optional and can be skipped if the preceding setup has been done. To do the setup, run the following commands in the given order:

```
git clone https://github.com/appium/appium.git
cd appium
npm install
gulp transpile # requires gulp, see below
npm install -g authorize-ios # for ios automation
authorize-ios # for ios automation
node
```

Selecting IDE

For the purpose of test development, we will choose **IntelliJ** (https://www.jetbrains.com/idea/) as the preferred IDE. Download the community edition from https://www.jetbrains.com/idea/download/. Once downloaded, open the .dmg package and drag IntelliJ to the Applications folder.

App under test

We will be using the Quikr app (Google Play link for app: https://play.google.com/store/apps/details?id=com.quikr&hl=en) throughout the book for Appium concepts and demonstration. All features of Appium can be demonstrated using this app. We are using the Quikr android app and mobile web version. It's easy to relate to any classifieds app, and it let us use gestures as well. For your learning, you can use any app of your choice which you are comfortable with. Book is written and evolved in such a way that it demonstrates the appium and automation concepts that can be applied to any app.

Machine setup for Windows

Machine setup for Windows will be a little different from that of Mac as we don't have the concept of a package manager. We will need to download the individual installers and run them to install the software we need. Let's start with installing Java, Android SDK, and then appium.

Installing Java

Following are the steps to install Java:

1. Visit the JDK download page and download the (`jdk-8uversion-windows-xxx.exe`) package based on your machine configuration (either the amd64 or x64).

2. Install Java from the downloaded package.

3. Once installed, bring up the search box and type `advanced system setting`. Click on the **View advanced system settings** search result.

4. On the system properties window, click on the **Advanced** tab and click on **Environment Variables**.

5. Under the **System variables** section, click on **New** and add a variable name--**JAVA_HOME**--and check for the installed location of the JDK. It will be similar to `C:/Program Files/Java/jdk1.8.xxx`.

6. Under the **System variables** section, scroll to find **PATH** and click on **edit**. Add `%JAVA_HOME%\bin` at the end.

7. Once done, launch the Command Prompt and type `java -version`; you should see the illustrated output with different version details based on the JDK version you installed:

```
java version "1.8.0_73"
Java(TM) SE Runtime Environment (build 1.8.0_73-b02)
Java HotSpot(TM) 64-Bit Server VM (build 25.73-b02, mixed mode)
```

8. You can also try running the `echo %JAVA_HOME%` command in Command Prompt, which should display the path we set earlier.

Installing Android SDK (using Android command-line tool)

1. Navigate to Android Studio page and download the Android Studio package (**android-studio-bundle-xxx.xxxxx-windows.exe**) for Windows (Link for Android Studio: `https://developer.android.com/studio/index.html?hl=sk`).

2. Once downloaded, run the `.exe` and follow the install instructions:

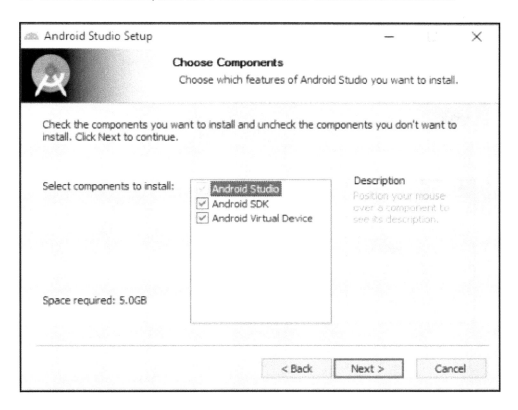

3. Click on **Next >**.
4. Click on the **I Agree** button.
5. Create a folder in C drive and name it `android-sdk`.

6. On the **Configuration Settings Install Locations** screen, choose the shown values for Android Studio and Android SDK:

7. Click on **Next**.
8. Finish the installation.
9. Bring up the Windows program search box and type `advanced system setting`. Click on the **View advanced system settings** search result.
10. On the system properties window, click on the **Advanced** tab and click on **Environment Variables**.
11. Under the **System variables** section, click on **New** and add a variable name, `ANDROID_HOME`, and value, `C:\Android-SDK`.
12. Under the **System variables** section, scroll to find **PATH** and click on **edit**. Add `%ANDROID_HOME%\tools` and `%ANDROID_HOME%\platform-tools` at the end.

13. To create an Android Virtual Device, we need to follow these steps:
    1. Launch **Android Studio** from the installed programs on Windows.
    2. On the **Android Studio** home page, click on the **Configure** dropdown and select **SDK Manager**.
    3. Under **SDK platform**, choose the SDK platform you want to install, such as **Android 5.1 (Lollipop)**, and select the checkbox.
    4. Click on **Apply** and confirm the installation.
    5. This will finish the installation of the new virtual device.
    6. Click on **OK** to close the popup.

14. Follow the instructions here to set up the AVD on a Windows machine (`https://developer.android.com/studio/run/managing-avds.html`).

Once done, we can move on to the Node JS installation.

Installing Node JS

Before we install Appium on Windows, we need to install Node JS. Navigate to the web page (`https://nodejs.org/en/download/`) and download the windows installer (`xxx.msi`) based on your architecture (either 32-bit or 64-bit).

Once downloaded, install the same with the default options. The `npm` and `nodejs` paths should be in your PATH environment variable.

Installing Appium

1. Download the `AppiumForWindows.zip` file from the location by visiting this link: `https://bitbucket.org/appium/appium.app/downloads/`.
2. Install the downloaded file and proceed with the default selections.
3. Launch the Appium app; it will open the permission popup for Node JS and allow that.

Installing Appium server (via npm)

1. Launch Command Prompt (Use Run as Administrator option) and type in this command:

   ```
   npm install -g appium
   ```

2. Once the preceding command is done, type in the next one:

   ```
   npm install -g appium-doctor
   ```

3. Once done, run the following command:

   ```
   appium-doctor
   ```

This will show the given output:

```
 Select C:\WINDOWS\system32\cmd.exe

C:\>appium-doctor
info AppiumDoctor  Appium Doctor v.1.2.5
info AppiumDoctor  ### Diagnostic starting ###
info AppiumDoctor   ANDROID_HOME is set to: C:\Android SDK
info AppiumDoctor   JAVA_HOME is set to: C:\Program Files\Java\jdk1.8.0_111
info AppiumDoctor   adb exists at: C:\Android SDK\platform-tools\adb.exe
info AppiumDoctor   android exists at: C:\Android SDK\tools\android.bat
info AppiumDoctor   emulator exists at: C:\Android SDK\tools\emulator.exe
info AppiumDoctor   Bin directory of $JAVA_HOME is set
info AppiumDoctor  ### Diagnostic completed, no fix needed. ###
info AppiumDoctor
info AppiumDoctor  Everything looks good, bye!
info AppiumDoctor

C:\>
```

Installing Genymotion

Sign up for a Genymotion account and download the windows installer with virtual box. Installing Genymotion on Windows is fairly simple; follow the default settings to proceed. The steps for creating the Genymotion emulator will remain the same as described earlier.

Selecting IDE

For test development purposes, we will choose IntelliJ (`https://www.jetbrains.com/idea /`) as the preferred IDE. Download the community edition from `https://www.jetbrains.c om/idea/download/`. Once downloaded, open the `.exe` package and follow the steps on the installation popup.

Appium GUI app

Let's take a detailed look at the Appium GUI app. Here's a snapshot of the Appium GUI app (on Mac OSX), which has a couple of icons on top, a console window, and a trash bin icon at the bottom. Windows Appium app has couple of options lesser than the Mac OS X, however functionally it represent the same as described below. Let's take a look at each of the icons and what it helps us to do:

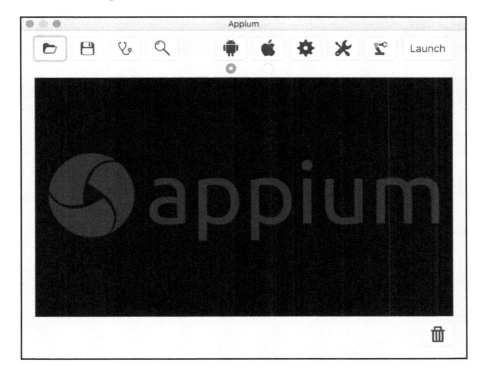

- **Open Configuration**: This lets you select any previously saved Appium configuration and load.

- **Save Configuration**: This saves the current Appium config, which is basically the settings you selected with Android or iOS or General.
- **Doctor**: It runs the `appium-doctor.js` program and tells you whether all the wirings are proper. By wiring, I mean it performs the following checks:
    - Xcode path
    - Xcode command-line tool
    - Checks for DevToolsSecurity to be enabled
    - Node.js installation
    - **ANDROID_HOME** to be set
    - **JAVA_HOME** to be set
- **Inspector**: It brings up the Appium inspector window. With the recent version of Appium, it launches the app on the device or emulator and shows the captured UI state of the application. The following is a snapshot of the same. The panel on the extreme right is clickable. Once you click on any UI element, the panel on the left shows the UI hierarchy and the panel named **Details** shows the UI attribute of the element clicked on. The right-hand side of the panel lets you click on any element that you need to interact with as part of your test code. On the **Details** pane, you can find attributes such as **type**, **text**, **index**, and **resource-id**:

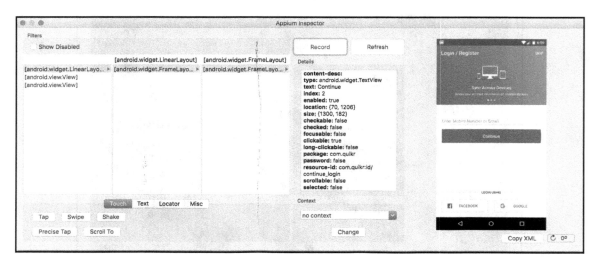

On the top of the Appium Inspector, you find the **Record** button, clicking on which generates the boiler plate code; it records the action on the element, based on the XPath of the element. Clicking on **Refresh** refreshes the panel on the left-hand side to load the latest UI snapshot on the right of the inspector screen and reloads the UI hierarchy and details panel as well.

Android Settings: Clicking on this brings up a window that lets you fill the android app-related details. The following is its snapshot. It is divided into two parts: **Basic** and **Advanced**. Under **Basic**, there are some mandatory settings and some optional ones:

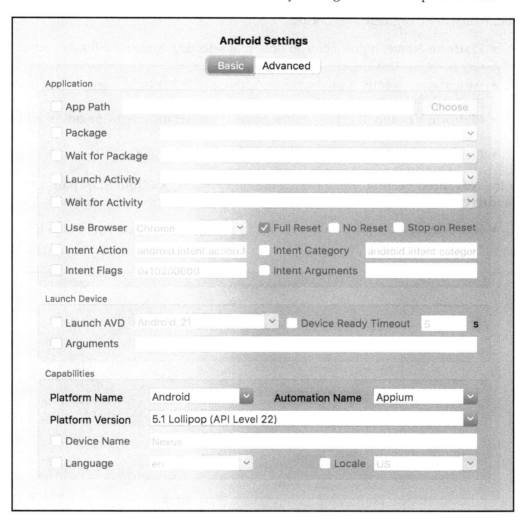

To launch an Appium session for a particular app, we need the app to be installed on the device, in which case we need the **Package** name and **Launch Activity.** If we want to install the app on the device and then start the Appium session, we need to pass in the **App Path** parameter, which is the location of the apk. So, we need to pass either the **App Path** or the **Package** and **Launch Activity** for the application section.

For the **Launch Device** section, **Launch AVD** will work if you have android emulator created via Android SDK. It doesn't work with Genymotion Emulator.

For the **Capabilities** section, there are four mandatory parameters, described here, and two optional, which are **Language** and **Locale**:

- **Platform Name**: It gives you an option of selecting Android if it's an android device or FirefoxOS.
- **Automation Name**: It can be either Appium or Selendroid. For this book, we will primarily be using Appium.
- **Platform Version**: It gives you the option to select different versions of android (such as API 22 and 21).
- **Device Name**: It is mandatory again and can be any text.

Under the **Advanced Section**, you have the option to choose **Android SDK path**, **Chromedriver Path**, and any **Keystore** settings. All these are optional parameters.

- **iOS Settings**: Clicking on this brings up a window that lets you fill the iOS app-related details:

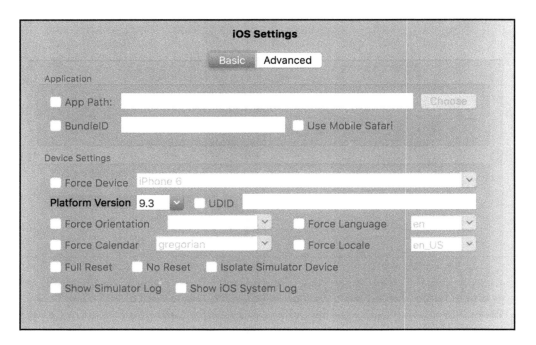

Before we proceed with any steps on iOS app automation using Appium, one of the important prerequisites is that the app must be signed with a developer identity:

- **App Path**: If you want to deploy the app using Appium, enter the location of .ipa (for physical device) or .app (for simulators).
- **BundleID**: If you want to invoke the existing app, then pass the **Bundle ID**.
- **Platform Version**: This lets you select the version of iOS that you want to connect Appium with.
- **UDID**: This is needed when you want to run the test on a physical device.
- **Force Device**: This lets you select the device on which you want the test to connect.
- **Full Reset**: This tells Appium to reset the state of the application we are testing every time we run a test.

On the **Advanced** tab, you need to be sure that Appium knows the path of Xcode on your machine. You can even change the path to Xcode using the **Change** button.

In the bottom section of the screen, you can see the buttons highlighted, such as **Touch**, **Text**, **Locator**, and **Misc**.

Clicking on **Touch** gives you an option, such as **Tap**, **Swipe**, and **Shake**, to be performed on the element you have selected on the right. Similarly, when you click on **Text**, it brings up a textbox to send in the text you want:

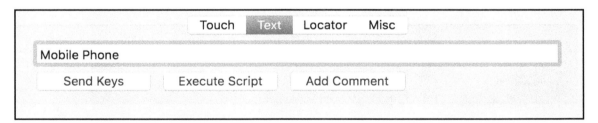

When you click on **Locator**, it brings up the option to choose the strategy with which you want to select the element, as illustrated in the following screenshot. The options available as part of strategy are **accessibility id**, **android uiautomator**, **class name**, **id**, **ios uiautomation**, **name**, and **xpath**. These are the different ways of identifying a UI element on a mobile app:

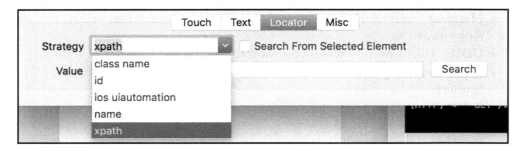

For example, in the following screenshot, the **Continue** button (on the right panel) on the app can be identified in two ways:

- Using **id**, we can use the ID as strategy and give the `continue_login` value that is taken from **resource-id**, highlighted in blue in the **Details** panel.
- Using **xpath**, we can construct a meaningful xpath that can be derived from the type of element and some unique attribute, which is **text** in this case. Hence, the xpath for the **Continue** button can be `//android.widget.TextView[@text="Continue"]`, same is illustrated below in the Appium inspector snapshot:

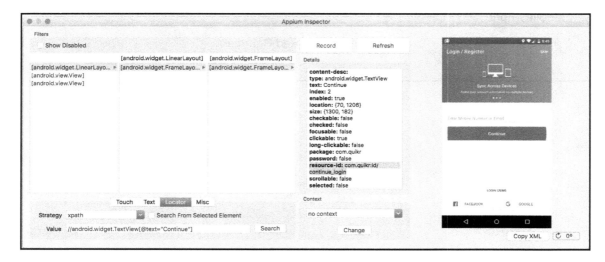

When you click on **Misc,** it gives you a button that performs the function of accepting or dismissing alerts.

Context: Following is a small section of the **Details** panel, called **Context**. It's a dropdown that shows you the context available for the app, be it native or web view. The next screenshot shows that the contexts available are both the native and web view. Appium lets you switch the context within the web driver protocol itself so that testing the native and web view parts becomes seamless. So, we can select one of the values in the dropdown and change the context:

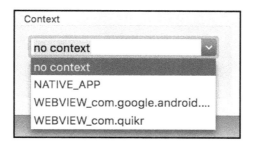

Summary

In this chapter, we learned to set up the machine both for Windows and Mac. We also learned to install Android SDK and update the system path for the same. We learned how to configure an Android Virtual Device and also explore the Genymotion emulator. We also learned to set up Appium, both the server and the GUI app. We explored the Appium GUI app and learned about the different settings we need to use, both for the android and iOS apps. We also explored how we can look up UI locator and check for the context in the app.

In the next chapter, we will start with setting up a project and write the first Appium test.

3
Writing Your First Appium Test

In the last two chapters, we saw what Appium is and how to set up the machine for both Mac OSX and Windows. Now that we have the ecosystem set up, let's start using Appium and writing some actual tests on Appium. In this chapter, we will set up an Appium Java project using IntelliJ and write our first test.

So, the set up we need before we actually write the code is this:

- Create a sample Java project
- Add Appium (automation tool) as a dependency
- Add Cucumber-JVM as a dependency
- Write a small test for a mobile web

For this example project, we will use **Cucumber** to write the specification. Cucumber is a tool based on the behavior-driven development framework. We have a separate section in this chapter that briefly talks about Cucumber.

 While authoring this book, both the Mac OSX as well as Windows machine support has been included. However, the features of Appium on Windows is not in sync with Mac OS Appium app. Windows users might find the Appium UI option to be missing on certain screens, please do proceed with the most similar option available.

Creating an Appium Java project (using gradle)

Let's create a sample Appium Java project in IntelliJ. This forms the foundation of all the code-related and Appium-related discussions we will have in the subsequent chapters. The following steps help you to achieve this:

1. Launch **IntelliJ** and click on **Create New Project** on the welcome screen.
2. On the **New Project** screen, select **Gradle** from the left pane. Project SDK should get populated with the Java version.
3. Click on **Next**, enter the **GroupId** as `com.test` and **ArtifactId** as `HelloAppium`. The version will already be populated; click on **Next**.
4. Check the **Use auto-import** option and ensure that **Gradle JVM** is populated. Click on **Next**. In case the Gradle JVM is not populated, please follow the below steps:

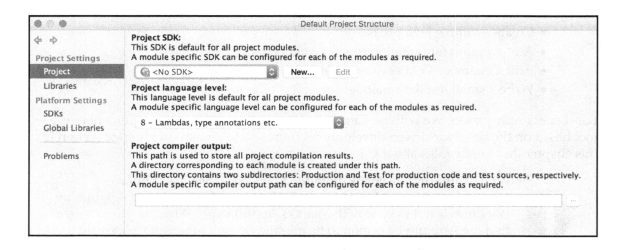

1. Click on **Configure** > **Project Defaults** > **Project Structure**:

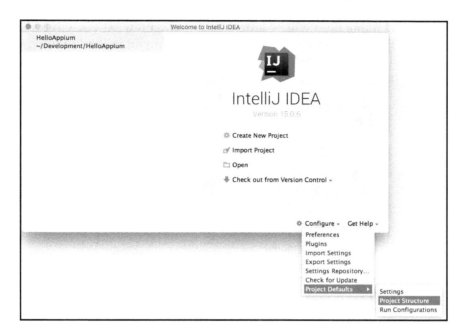

2. Choose **Project** under **Project Settings** as shown below:

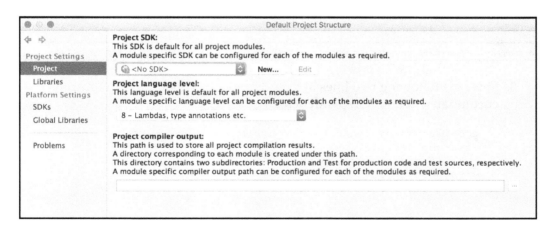

3. Click on **New...** button.
4. Point it to the JDK installed on your machine.

5. Click on OK to close the pop up and go to the new Project creation screen.

5. The **Project name** field will be auto-populated with what you gave as **ArtifactId**. Choose a **Project location** and click on **Finish**. IntelliJ will be running the background task (Gradle build), which can be seen in the status bar.

6. This should create a project with the following structure:

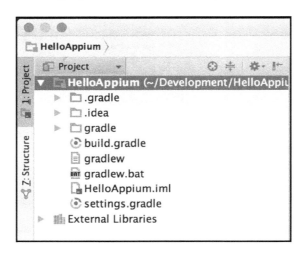

7. Open the `build.gradle` file. You will see a message, as shown; click on Ok, apply suggestion!:

You can configure Gradle wrapper to use distribution with sources. It will provide IDE with Gradle API/DSL... Hide the tip   Ok, apply suggestion!

8. Enter the following two lines in `build.gradle`. This adds Appium and **cucumber-jvm** under **dependencies**:

```
compile group: 'info.cukes', name: 'cucumber-java',
version: '1.2.5'
compile group: 'io.appium', name: 'java-client',
version: '5.0.0-BETA6'
```

9. Here's how the gradle file should look:

```
group 'com.test'
version '1.0-SNAPSHOT'

apply plugin: 'java'

//sourceCompatibility = 1.8

repositories {
    mavenCentral()
}

dependencies {
    testCompile group: 'junit', name: 'junit', version: '4.11'
    compile group: 'info.cukes', name: 'cucumber-java', version: '1.2.5'
    compile group: 'io.appium', name: 'java-client', version: '5.0.0-BETA6'
}
```

10. Once done, navigate to **View** > **Tools Windows** > **Gradle** and click on the Refresh all gradle projects icon. This will pull all the dependencies in External Libraries:

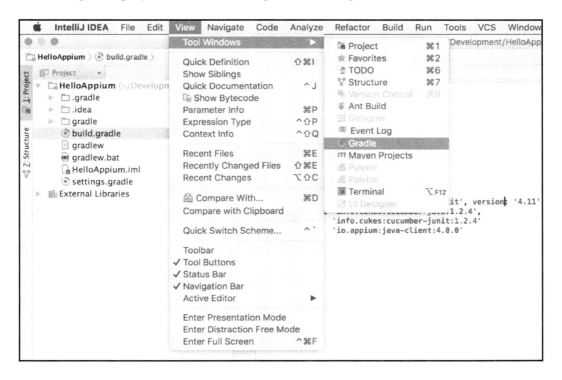

11. Navigate to **Preferences** > **Plugins**, search for **Cucumber for Java**, and click on **Install** (if it's not previously installed).

12. Repeat the preceding step for **Gherkin** and install the same. Once done, restart IntelliJ if it prompts.

Once done, we are ready with the IntelliJ project creation. The next step is to write a Cucumber feature file; however, let's first understand what Cucumber is.

Introduction to Cucumber

Cucumber is a test framework that supports behavior-driven development (or BDD, in short). The core idea behind BDD is domain-specific language (known as DSL) where the tests are written in normal English, expressing how the application or system has to behave. DSL is an executable test that starts with a known state, performs some action, and verifies the expected state:

```
Given I launch the app
And I click on Register
Then I should see register with Facebook and Google
```

 Dan North (creator of BDD) defined behavior-driven development in 2009 as --BDD is a second-generation, outside-in, pull-based, multiple-stakeholder, multiple-scale, high-automation, agile methodology. It describes a cycle of interactions with well-defined outputs, resulting in the delivery of working, tested software that matters.

Cucumber feature files serve as a living documentation that can be implemented in many languages. It was first implemented in Ruby and later extended to Java. Some of the basic features of Cucumber are listed as follows:

- The core of Cucumber is text files called features, which contain scenarios. These scenarios express the system or application behavior.
- Scenario files consist of steps that are written following the syntax of Gherkin.

A sample feature file is as shown here:

```
Feature: Sign up

Scenario: Facebook Integration
A user should be able to register and log into the app by using the
Facebook account.

Given I choose to sign up
And I select to sign up using Facebook
Then I should see a pop up with my Facebook account to continue
And I should be logged in once I allow
```

So, in the preceding example, `Feature`, `Scenario`, `Given`, `But`, `Then`, and `And` are keywords. Let's take a look at some of the most used keywords of Cucumber and what it means:

Feature: tests are grouped into features. We use this name because we want engineers to describe the features that a user will be able to use.

Scenario: A scenario expresses the behavior we want. Each feature contains several scenarios; each scenario is a example of how the system should behave in a particular situation. The expected behavior of the feature will be the total scenarios. For a feature to pass, all scenarios must pass.

Test Runner: There are different ways to run the feature file; however, we will use the JUnit runner initially and then move on to the gradle command for command-line execution.

So, I am hoping that we now have a brief idea of what Cucumber is. Further details can be read on their site (`https://cucumber.io/`). In the following section, we will create a feature file, write a scenario, implement the code behind, and execute it.

Writing our first Appium test

Until now, we have created a sample Java project and added the Appium dependency. Next, we need to add a feature file and implement the code behind. Let's start that:

1. Under the `Project` folder, create the `src/test/java/features` directory structure.
2. Right click on the `features` folder, select **New > File,** and enter name as `Sample.feature`.

3. You will notice that the file is associated with a Cucumber feature icon if the plugin is installed correctly.
4. We need to explore the Quikr mobile app; when you install it and play around the first scenario, you will notice the login scenario. Quikr gives you an option to log in using Google or Facebook.
5. In the `Sample.feature` file, let's write a sample scenario, as shown, which is about logging in using Google.
6. Detailed steps will be clicking on log in using Google, and then verifying that the account picker screen has a valid email ID:

```
Feature: Hello World

Scenario: Registration Flow Validation via App
As a user I should be able to see my google account
when I try to register myself in Quikr

When I launch Quikr app
And I choose to log in using Google
Then I see account picker screen with my email address
"testemail@gmail.com"
```

7. When the Cucumber steps are not implemented, it will highlight them in yellow. Right now, all the steps will be highlighted. The implementation of these steps will be java class, and they can be hosted under different packages.
8. Right-click on the `java` folder, select **New > Package**, and enter name as `steps`.
9. The next step is to implement the Cucumber steps; click on the first line in the `Sample.feature` file `When I launch Quikr app` and press *Alt+Enter*. Then, select the Create step definition option:

```
Feature: Hello World

  Scenario: Registration Flow Validation via app
  As a User I should be able to see my google account
  when I try to register myself in Quikr app

    When I launch Quikr app
    And I choose to log in     ● Create step definition    ▶  s "testemail@gmail1.com"
    Then I see account pick    💡 Create all steps definition ▶
```

10. It will present you with a popup to enter **File name**, **File location**, and **File type**. We need to enter the step's class name; select the shown values. Since the step belongs to Home Page (or we can even call it Landing Page), we create the `HomePageSteps` class:

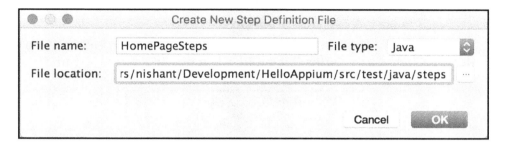

11. So, the idea is that the steps will belong to a page and each page will typically have its own step implementation class.

12. Once you click on **OK**, it will create the given template in the `HomePageSteps` class file:

```
public class HomePageSteps {
@When("^I launch Quikr app$")
public void iLaunchQuikrApp() throws Throwable {
// Write code here that turns the phrase above into
concrete actions
throw new PendingException();
}
}
```

We have written our test steps; the next step is to implement the actual code behind which will launch the Appium server and then deploy the app on the emulator. So, let's start with launching the emulator first and then the Appium App (For Windows user, the screen below may be totally different and some options may not be present; you can choose the mandatory options and ignore others if it is not present):

1. Download the Quikr app (version 9.16).
2. Create a folder named `app` under the `HelloAppium` project and copy the downloaded `apk` under that folder. Ideally, this will host the app under test:

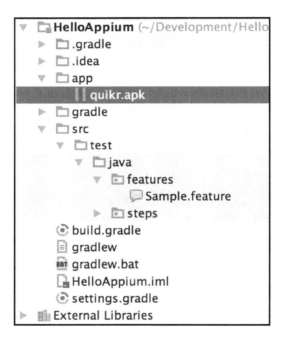

3. Launch the Appium GUI app:

- On Mac, navigate to **Finder** > **Applications** > **Appium**.
- On Windows, navigate to **Start Menu >** Type **Appium** > **Press Enter**:

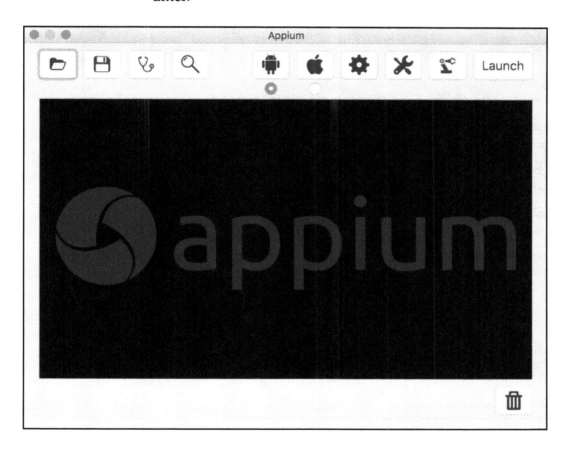

4. Launch the Genymotion emulator (the one we created in Chapter 2, *Setting Up the Machine*) by selecting the virtual device and clicking on the **Start** icon, as highlighted; wait for it to get started:

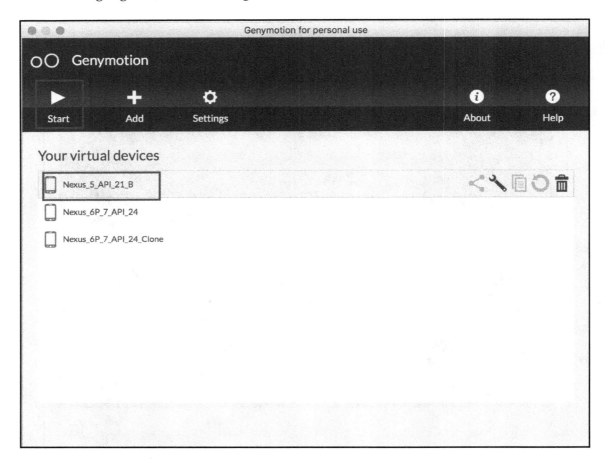

5. Once launched, we will see the Genymotion emulator, as shown:

6. Switch to the Appium GUI app, click on the android icon, and enter the following details:

- App Path: Browse to the .apk location under the app folder
- Platform Name: Select Android
- Automation Name: Select Appium
- Platform Version: Select 5.1 Lollipop (API Level 22) from the dropdown, as the emulator created in `Chapter 2`, *Setting Up the Machine* or the device. Also, even though it's a dropdown, it allows you to edit the value and it behaves as a text input field rather than a dropdown. On Windows app , it's just a drop down so make sure you have a emulator version which is supported by the Appium windows app.

Android N (version 7.0) has support issues with Appium. Be cautious while you are trying your hands with version 7.0.

- **Device Name**: Enter any string, such as **Nexus6**:

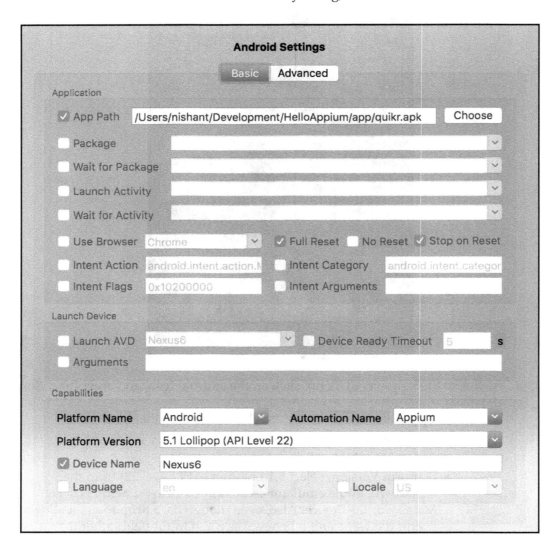

7. Once the preceding settings are done, click on the **General Settings** icon-- ✿ --and choose the following settings:

- Select **Pre-Launch Application**.
- Select **Strict Capabilities**.
- Select **Override Existing Sessions**.
- Select **Kill Processes Using Server Port Before Launch**.
- Select New Command Timeout and enter the value **7200**. Refer to the following screenshot:

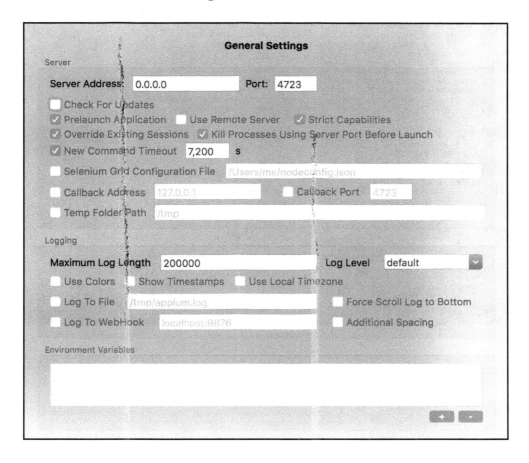

8. Once done, close the popup by clicking on the General settings icon-- ✿ --again. Then, click on Launch.

9. You will see a bunch of logs on the Appium console and will be able to find the given line:

```
[Appium] Appium REST http interface listener started on
0.0.0.0:4723
[HTTP] --> GET /wd/hub/status {}
[MJSONWP] Calling AppiumDriver.getStatus() with args: []
[MJSONWP] Responding to client with driver.getStatus() result:
{"build":{"version":"1.5.3"...
[HTTP] <-- GET /wd/hub/status 200 17 ms - 83
```

So basically, it indicates that the Appium server has been started on the default server and port, and it has returned the status **HTTP 200**:

1. Click on the **Inspector** icon-- ⌕ --to launch the Appium Inspector popup.
2. The preceding step will install the app on the emulator and launch the inspector window. It will also install the Appium Settings app and Unlock app on the emulator.
3. If the emulator image in the right pane of the Inspector is not fully loaded, then click on the **Refresh** button to sync with the t on the emulator. Here's how the Inspector popup will look after clicking on the **Refresh** button:

If you notice the Appium console, you will find the following log when you click on **Inspector**:

```
[HTTP] --> POST /wd/hub/session
{"desiredCapabilities":{"platformName":"Android","platformVersion":"5.1","n
ewCommandTimeout":"7200","app":"/Users/nishant/Development/HelloAppium/app/
quikr.apk","automationName":"Appium","deviceName":"Nexus6"}}

[MJSONWP] Calling AppiumDriver.createSession() with args:
[{"platformName":"Android",...
```

```
[Appium] Creating new AndroidDriver session
[Appium] Capabilities:
[Appium] platformName: 'Android'
[Appium] platformVersion: '5.1'
[Appium] newCommandTimeout: '7200'
[Appium] app:
'/Users/nishant/Development/HelloAppium/app/quikr.apk'
[Appium] automationName: 'Appium'
[Appium] deviceName: 'Nexus6'
[Appium] fullReset: true
[BaseDriver] Capability 'newCommandTimeout' changed from string
('7200') to integer (7200). This may cause unexpected behavior
[BaseDriver] Session created with session id: 6c61a910-5c7e-
44ff-9cb1-fb2413805cda
[debug] [AndroidDriver] Getting Java version

[AndroidDriver] Java version is: 1.8.0_73

[ADB] Checking whether adb is present

[ADB] Using adb from /usr/local/Cellar/android-
sdk/24.4.1_1/platform-tools/adb
[AndroidDriver] Retrieving device list
[debug] [ADB] Trying to find a connected android device
[debug] [ADB] Getting connected devices...
[debug] [ADB] 1 device(s) connected
[AndroidDriver] Looking for a device with Android 5.1
[debug] [ADB] Setting device id to 192.168.56.101:5555

[ADB] Getting device platform version

[debug] [ADB] Getting connected devices...
```

The last line of the preceding log shows that the Appium server has created a session with session ID `6c61a910-5c7e-44ff-9cb1-fb2413805cda`. For all future communication with the device until the server is alive, this session ID will be the context. For example, when you click on the **Refresh** button on the inspector, the console will log the request like `[HTTP] --> GET /wd/hub/session/6c61a910-5c7e-44ff-9cb1-fb2413805cda/source {}`.

The session gets killed when we click on **Stop** on the Appium GUI. So every time we click on **Start,** Appium gives us a new session; this is because we selected the parameter **Override ExistingSession**. This makes sure that the previous Appium session is killed before creating a new one.

Now Appium has started the server and created a session based on the parameters we passed in both the **Android Settings** and **General Settings**, which are also known as **Desired Capabilities**. We will take a detailed look at **Desired Capabilities** in the next chapter.

1. Coming back to automation of the steps, we need to implement whatever we have done until now as code.
2. Click on the **Record** button on **Appium Inspector**; it will generate the boilerplate code, which will perform the function of setting device-related capabilities (such as Platform name and version) when the Appium server is already running:

Windows machine do not have the **Record** button. Windows users can skip to step 3 and copy the code mentioned below. The code generated is OS independent, it will work seamlessly for both Mac OSX and Windows user.

3. Copy the preceding code (generated by appium) and paste the same in the `iLaunchQuikrApp` method of the `HomePageSteps` class. Delete the line `throw new PendingException();`.

4. Resolve the dependencies by importing the necessary class or add the following code to the class file:

```
import io.appium.java_client.AppiumDriver;
import org.openqa.selenium.remote.DesiredCapabilities;
```

5. Remove the `wd.close` line as this will kill the session. We need the session to perform other tests, and then kill the session once done.

6. This is how the code snippet will look:

```
@When("^I launch Quikr app$")
public void iLaunchQuikrApp() throws Throwable {
DesiredCapabilities capabilities = new DesiredCapabilities();
capabilities.setCapability("appium-version", "1.0");
capabilities.setCapability("platformName", "Android");
capabilities.setCapability("platformVersion", "5.1");
capabilities.setCapability("deviceName", "Nexus6");
capabilities.setCapability("app",
"/Users/nishant/Development/HelloAppium/app/quikr.apk");
AppiumDriver wd = new AppiumDriver(new
URL("http://0.0.0.0:4723/wd/hub"), capabilities);
wd.manage().timeouts().implicitlyWait(60, TimeUnit.SECONDS);
}
```

7. So, the boilerplate typically has all the settings we made under **Android Settings**.

8. Close the Appium Inspector window and click on **Stop** on the Appium GUI app. This kills the current Appium session.

We will discuss some of the concepts, such as and implicit wait, in the upcoming chapters. To create an Appium session we need only 4 capabilities to be passed which is generated by the boiler plate code: `platformName`, `platformVersion`, `deviceName` and `app`.

Let's try to run the generated code to see whether it works seamlessly.

Running the feature file

With the Appium GUI app, we did two things:

1. Ran the Appium server
2. Set the desired capabilities

To run the preceding code from IntelliJ, we need Appium server to be running. From the preceding code, we are only setting desired capability and getting a session of our choice, but the server has to be running:

1. The Appium server can be started either by the Appium GUI app or the Appium command line. Let's use a command line for test execution purpose.

2. Launch the terminal (Command Prompt in case of Windows machine), type in the appium command, and press **Enter** to run the Appium server. We can see the Appium logs in the console.

```
? ~ appium
[Appium] Welcome to Appium v1.6.3
[Appium] Appium REST http interface listener started on
0.0.0.0:4723
```

3. In **IntelliJ**, right-click on the scenario file and select the **Run 'Scenario:Registration Flow...'** option.

4. This will be the sample output:

```
Testing started at 9:33 AM ...
Jan 15, 2017 9:33:34 AM
org.openqa.selenium.remote.ProtocolHandshake
createSession
INFO: Attempting bi-dialect session, assuming Postel's Law holds
true on
the remote end
Jan 15, 2017 9:33:39 AM
org.openqa.selenium.remote.ProtocolHandshake
createSession
INFO: Falling back to original OSS JSON Wire Protocol.
Jan 15, 2017 9:34:54 AM
org.openqa.selenium.remote.ProtocolHandshake
createSession
INFO: Detected dialect: OSS

Undefined step: And I choose to log in using Google

Undefined step: Then I see account picker screen with my email
address
"test@gmail.com"

1 Scenarios (1 undefined)
3 Steps (2 undefined, 1 passed)
1m21.575s
```

5. So, the first step has passed and the other two are yet to be implemented; we are good to proceed with automating the rest of the steps.

6. Kill the appium server once the test is run. Navigate to Terminal (Command Prompt on Windows) and press *Ctrl + C* to kill the process.

As and when we progress, we will keep refactoring the code to make it more readable and maintainable.

Refactoring

In the preceding code we wrote, `AppiumDriver wd` is private to the method. We need to refactor it to make it available to all the methods in that class. So, highlight the line, click on **Refactor > Extract > Field**, and choose `appiumDriver` from the list of values shown.

This will make your code look like this:

```
public class HomePageSteps {

    private AppiumDriver wd;

    @When("^I launch Quikr app$")
    public void iLaunchQuikrApp() throws Throwable {
        DesiredCapabilities capabilities = new DesiredCapabilities();
        capabilities.setCapability("appium-version", "1.0");
        capabilities.setCapability("platformName", "Android");
        private AppiumDriver wd;        "platformVersion", "5.0");
                                        "deviceName", "Nexus6");
        capabilities.setCapability("app", "/Users/nishant/Development/HelloAppium/app/quikr.apk");
        AppiumDriver wd = new AppiumDriver(new URL("http://0.0.0.0:4723/wd/hub"), capabilities);
        wd.manage().  wd                              t.SECONDS);
    }                  appiumDriver
                       driver
    @And("^I choose  Press ⌥⌘F to show dialog with more options
    public void iChooseToLoginUsingGoogle() throws Throwable {

    }
```

Implementing the remaining steps

Come back to the `Sample.feature` file and create step definitions for the other two steps in the sample class file, **HomePageSteps**.

Step 2 implementation:

1. In the `And I choose to log in using Google` step, we are supposed to click on the Google icon once the app launches.

2. The sequence of steps is:
     1. Find the locator for the Google icon.
     2. Click on the icon.

3. To find the locator, we need the Appium GUI app (Since the android settings are correct and unaltered, we can start the session again by clicking on **Launch**).

4. Once the server has started, click on the Appium inspector icon and wait for the app to launch on emulator.

5. Click on the **Refresh** button in the Appium Inspector popup.

6. Once done, click on the **Google icon** in the Appium Inspector right panel; it will show you the layout details and button attributes in the **Details** panel. The *id* for the Google button is highlighted in red with row name **resource-id** in the **Details** section. The first part there--`com.quikr`--is the **package name**; the value of **id** is `sign_in_button`. The inspector also gives you the xpath, which is the last item under the **Details** section. There are different ways to find an element: `id`, `className`, `cssSelector`, `linkText`, `partialLinkText`, `name`, `tagName`, `xpath`. We can choose one of these based on what is available. We will see locators in detail in `Chapter 5`, *Understanding Appium Inspector to Find r*:

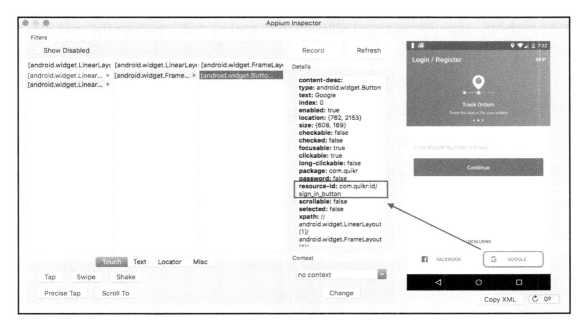

7. For now, we can use the **id** field. Once we get the value of id, we tell the r to find an element by the property name `id` and value `sign_in_button`; then, perform a click operation on it.

8. Navigate to the `iChooseToLogInUsingGoogle` method under the `HomePageSteps` class and paste the following code. Add an import statement `import org.openqa.selenium.By` for:

```
appiumDriver.findElement(By.id("sign_in_button")).click();
```

Step 3 implementation:

1. In the `Then I see account picker screen with my email address "test@gmail.com"` step, we are required to get the value of the email displayed on the Google account picker popup and match it with the expected value passed.

2. The sequence of steps is:
    1. Find the locator of the email ID field
    2. Get the value of that field
    3. Perform string comparison and pass or fail accordingly.

3. If you have stopped the Appium session, launch the Appium GUI app and wait for the app to be launched. Click on the Google icon and wait for the account picker to show up.

4. Once done, click on the Inspector icon on Appium GUI; here's how the Appium inspector screen will be:

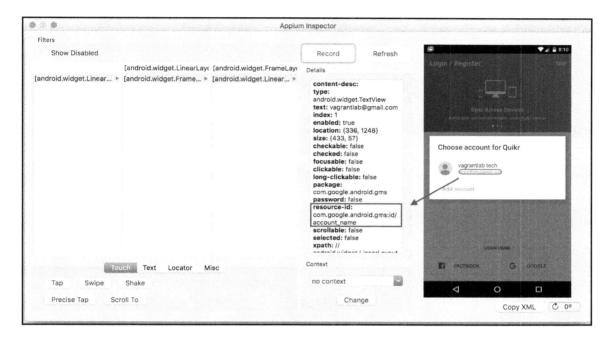

5. Let's quickly verify if using the ID value will give us that element. Click on Locator in the inspector screen. Select the strategy as **id**, enter value as `account_name`, and click on **Search**. It throws up an error, as illustrated:

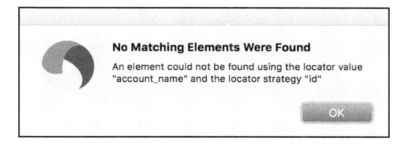

6. An important thing to notice in **resource-id** is that the package name has been changed to `com.google.android.gms`. Let's use the complete value of ID in this case--`com.google.android.gms:id/account_name`--and click on **Search**.

7. It works this time, showing one element found.

8. So, we are good to implement the `iSeeAccountPickerScreenWithMyEmailAddress` method, as follows. Rename the parameter from `arg0` to `expected` and copy and paste the following implementation. Since clicking on the Google button and showing up the popup will take time, we added a `Thread.sleep()` before the assert statement:

```
@Then("^I see account picker screen with my email address
\"([^\"]*)\"$")
public void iSeeAccountPickerScreenWithMyEmailAddress(String
expected) throws Throwable {
 Thread.sleep(5000);
 Assert.assertEquals("Email Id matches", expected,
appiumDriver.findElement(By.id("com.google.android.gms:id/account_n
ame")).getText());
 }
```

9. In the preceding method, we are using JUnit assertion--`Assert.assertEquals`--for string comparison. The advantage of this method is that it will show both the expected and the actual string in case of failure:

```
org.junit.ComparisonFailure: Email Id matches expected:
<vagrantlab@gmail[1].com> but was:<vagrantlab@gmail[].com>
at org.junit.Assert.assertEquals(Assert.java:115)
```

The complete `HomePageStep` class will look like this:

```
package steps;

import cucumber.api.java.en.And;
import cucumber.api.java.en.Then;
import cucumber.api.java.en.When;
import io.appium.java_client.AppiumDriver;
import org.junit.Assert;
import org.openqa.selenium.By;
import org.openqa.selenium.remote.DesiredCapabilities;

import java.net.URL;
import java.util.concurrent.TimeUnit;

public class HomePageSteps {
```

```
private AppiumDriver appiumDriver;

@When("^I launch Quikr app$")
public void iLaunchQuikrApp() throws Throwable {

DesiredCapabilities capabilities = new DesiredCapabilities();
capabilities.setCapability("platformName", "Android");
capabilities.setCapability("platformVersion", "5.0");
capabilities.setCapability("deviceName", "Nexus");
capabilities.setCapability("noReset", false);
capabilities.setCapability("fullReset", true);
capabilities.setCapability("app",
"/Users/nishant/Development/HelloAppium/app/quikr.apk");
System.out.println(capabilities.toString());
appiumDriver = new AppiumDriver(new
URL("http://0.0.0.0:4723/wd/hub"),
capabilities);
appiumDriver.manage().timeouts().implicitlyWait(60,
TimeUnit.SECONDS);
}

@And("^I choose to log in using Google$")
public void iChooseToLogInUsingGoogle() throws Throwable {
appiumDriver.findElement(By.id("sign_in_button")).click();
}

@Then("^I see account picker screen with my email address \"([^\"]*)\"$")
public void iSeeAccountPickerScreenWithMyEmailAddress(String expected)
throws Throwable {
Thread.sleep(5000);
Assert.assertEquals("Email Id matches", expected,
appiumDriver.findElement(By.id("com.google.android.gms:id/account_name")).g
etText());
}

}
```

Let's now move to the execution of our preceding code. Once we are done with the step implementation, stop the Appium GUI session.

Running the scenario

In the last section, we completed the implementation of the scenario. The next step is to execute the same. So, navigate to the feature file in IntelliJ and right-click on the scenario, which will show the illustrated option. We get two options: one is to run the scenario and the other is to debug the scenario:

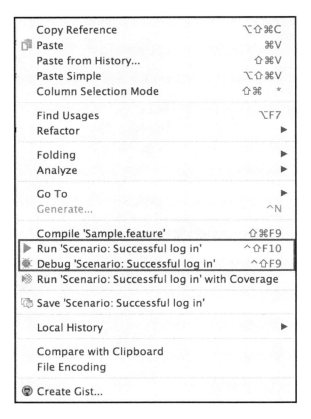

So, the steps to run the feature file are these:

1. Start the terminal (Command Prompt in the case of Windows Machine) and type in `appium --session-override`.
2. Choose the run scenario option (from the steps explained earlier).

The preceding command will launch the android emulator and start the test execution.

Automating a mobile web app using Appium

To automate a mobile web app, we need to have a mobile web browser installed on the emulator/device. All android phones generally come with a stock browser installed. For example, Genymotion comes with a stock browser installed and can be used for automation purposes. For emulators created using Android SDK, we need to install the Chrome browser by downloading its install file (.apk) based on the CPU configured for the emulator.

Another way to install the Chrome browser on the emulator is to install it via Play Store. To install Play Store on the emulator, you need to download the following files and install them first:

- com.android.vending-x.x.xx.apk
- com.google.android.gsf_x.x.x-xx_minAPIxx(nodpi).apk
- GoogleLoginService.apk

The files need to be for the Android version you created the emulator with. Once done, launch the Play Store and search for the Chrome app and download it.

Let's write another scenario to test the mobile web app of Quikr:

```
Scenario: Registration Flow Validation via web
As a User I want to verify that
I get the option of choosing Facebook when I choose to register

When I launch Quikr mobile web
And I choose to register
Then I should see an option to register using Facebook
```

Implement the first step and create the skeleton definition for the step in a separate class, HomePageWebSteps.Java. Here, we will not need the Desired Capability app as the test will be run on the browser on the device; instead, we will use browserName. The implementation is shown here:

```
@When("^I launch Quikr mobile web$")
public void iLaunchQuikrMobileWeb() throws Throwable {
  DesiredCapabilities desiredCapabilities = new
  DesiredCapabilities();
  desiredCapabilities.setCapability("platformName", "Android");
  desiredCapabilities.setCapability("deviceName", "Nexus");
  desiredCapabilities.setCapability("browserName", "Browser");

  URL url = new URL("http://127.0.0.1:4723/wd/hub");
  appiumDriver = new AppiumDriver(url, desiredCapabilities);
```

```
appiumDriver.get("http://m.quikr.com");
}
```

Compared to the hybrid app implementation explained earlier, the major difference lies with the desired capability `browserName` being used. This parameter tells the Appium about the browser being requested.

Possible values for `browserName` are as follows:

- **Chrome**: For a Chrome browser on Android<
- **Safari**: For a Safari browser on iOS
- **Browser**: For a stock browser on Android

To run the preceding step, follow the given steps:

1. Ensure that you have the emulator running and a stock browser installed. Genymotion emulator comes with one, so we can readily run the test.
2. Start the Appium server by running the `appium --session-override` command in the terminal (for Windows machine, use Command Prompt).
3. Right-click on the feature file and select the **Run scenario...** option for the last written scenario.

Implementing the remaining steps

To implement the remaining steps, we need to find locators for the elements we want to interact with. Once the locators are found, we need to perform the desired operation.

Following are the steps that will help us find the locators:

1. Launch the chrome browser and navigate to the mobile site (in our case, `http://m.quikr.com`).
2. Select **More Tools** > **Developer Tools** from the Chrome Menu.
3. In the Developer Tool menu items, click on the Toggle device toolbar icon, highlighted in blue:

4. Once done, the page will be displayed in a mobile layout format.
5. In order to find the locator of any UI element, click on the first icon of the dev toolbar and click on the desired element.

6. The HTML in the dev tool layout will change to highlight the selected element. Refer to this screenshot that shows the same:

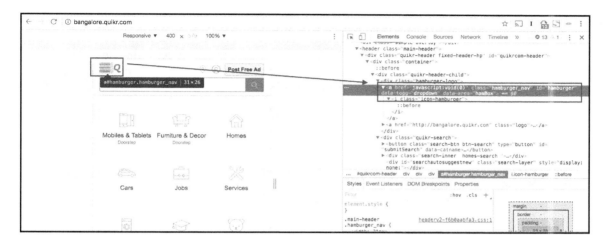

7. In the highlight panel on the right-hand side, we can see the `class=hamburger_nav` and `id=hamburger` properties. We can choose to use `id` and implement the step as follows:

```
appiumDriver.findElement(By.id("hamburger")).click();
```

Similarly, we can use the chrome dev tool to find the locator of the other element--**Sign In/Sign Up** link--and perform the click operation.

So, the second step can be implemented as shown:

```
@And("^I choose to register$")
public void iChooseToRegister() throws Throwable {
  appiumDriver.findElement(By.id("hamburger")).click();

  Thread.sleep(5 * 1000);
  appiumDriver.findElement(By.id("hamLoginLink")).click();

}
```

For now, we will use `Thread.sleep(long millis)` to wait for an action to complete and then refactor the same to implement using `WebDriverWait`. This will be explained thoroughly in a separate chapter.

The last method to verify the presence of Facebook as an option can be done in a couple of ways, one of them being asserting for the presence of an element and another way being asserting for certain text to be present in the t:

```
Then I should see an option to register using Facebook
```

So, to implement the preceding step, we can check whether the button with the name **Facebook** is displayed on the **Register** tab. Wherever there is a change in screen or we are expecting some element to appear or action to happen, we can use `Thread.sleep()` for w:

```
@Then("^I should see an option to register using Facebook$")
public void iShouldSeeAnOptionToRegisterUsingFacebook() throws Throwable {
  Thread.sleep(5 * 1000);
  appiumDriver.findElement(By.partialLinkText("Register")).click();
  Assert.assertTrue(appiumDriver.findElement(By.className("icon-facebook")).isDisplayed());

}
```

This completes writing the basic test of the mobile web app. So, basically, your class file-- `HomePageWebSteps`--should look like this:

```
public class HomePageWebSteps {
  private AppiumDriver appiumDriver;

  @When("^I launch Quikr mobile web$")
  public void iLaunchQuikrMobileWeb() throws Throwable {
  DesiredCapabilities desiredCapabilities = new
  DesiredCapabilities();
  desiredCapabilities.setCapability("platformName", "Android");
  desiredCapabilities.setCapability("deviceName", "Nexus");
  desiredCapabilities.setCapability("browserName", "Browser");

  URL url = new URL("http://127.0.0.1:4723/wd/hub");
  appiumDriver = new AppiumDriver(url, desiredCapabilities);
  appiumDriver.get("http://m.quikr.com");
  }

  @And("^I choose to register$")
  public void iChooseToRegister() throws Throwable {
  appiumDriver.findElement(By.id("hamburger")).click();

  Thread.sleep(5 * 1000);
  appiumDriver.findElement(By.id("hamLoginLink")).click();

  }
```

```
@Then("^I should see an option to register using Facebook$")
public void iShouldSeeAnOptionToRegisterUsingFacebook() throws
Throwable {
Thread.sleep(5 * 1000);
appiumDriver.findElement(By.partialLinkText("Register"))
.click();

Thread.sleep(2 * 1000);
Assert.assertTrue(appiumDriver.findElement(By.className("icon-
facebook")).isDisplayed());

}
}
```

Now, we can run the test scenario to check whether we get the option of registering using Facebook in the mobile web app.

Automating the iOS app using Appium

So far we have learned how to create a project, looked into Cucumber and how to write a feature file, took a sample android app and mobile web app to learn how to get started with writing our first test. Let's take a look at the iOS app now. Let's understand some dependency before we get started.

Some of the mandatory things we need:

- The iOS app automation needs Mac OS X (Windows user will not be able to execute the below steps)
- Xcode needs to be installed (Version 7.1)
    - In case you have installed the latest version of Xcode you will have to use the new Appium Desktop app.
    - You can also have multiple versions of Xcode on your machine and choose to default to the older version (less than 8) to run through the following example:

```
sudo xcode-select –switch <path/to/old/>Xcode.app
```

- App under test
    - The iOS simulator app (.app) is needed for any testing on iOS simulators. This is the debug version of the app.
    - The iOS app (.ipa) is needed for any testing on iOS real devices. To use this app for automation purpose, the app must be signed with a development identity (needs an iOS Developer license).
- For the reference in this book, we will use a simulator app which can be downloaded from Appium repo at: https://github.com/appium/ios-test-app

Build the app

Let's build the app for the simulator first which will form the basis of the whole chapter going forward and the discussions. Follow these steps:

1. Navigate to your local workspace or any folder of your choice
2. Clone the repo using the following command:

```
git clone https://github.com/appium/ios-test-app
```

3. It should show the following output once the command is successfully completed:

```
[➜ Development git clone https://github.com/appium/ios-test-app.git
Cloning into 'ios-test-app'...
remote: Counting objects: 378, done.
remote: Total 378 (delta 0), reused 0 (delta 0), pack-reused 378
Receiving objects: 100% (378/378), 80.63 KiB | 58.00 KiB/s, done.
Resolving deltas: 100% (183/183), done.
Checking connectivity... done.
[➜ Development cd ios-test-app
```

4. Navigate to the directory and build the repo using the following command:

```
npm install
```

5. Once the command is successfully executed, it will create a folder `build` as shown in the following screenshot, that contains the test app. Navigate to the file `build > Release-iphonesimulator > TestApp`:

```
[→ ios-test-app git:(master) ls
Default-568h@2x.png  TestApp.xcodeproj   gulpfile.js        lib
LICENSE              apps.json           index.js           node_modules
README.md            build               install-npm.js     package.json
Test App 2           build-js            install.js         test
[→ ios-test-app git:(master) cd build
[→ build git:(master) cd Release-iphonesimulator
[→ Release-iphonesimulator git:(master) ls
TestApp.app
→ Release-iphonesimulator git:(master) 
```

Deploying the app on the iOS Simulator

Once we have the app, we need to install the app on the iOS simulator. We have different options to do this:

1. Via `xcrun` command
2. By using Appium

Via xcrun command

To install app via the xcrun command we need to start the simulator first. Launch Xcode and launch the simulator by clicking **Xcode > Open Developer Tool > Simulator.**

Once the simulator is booted and running, start the terminal and run the command `xcrun simctl install booted path_to_the_app/TestApp.app` as shown. If the command is successful, it will return the prompt.

Using Appium

Let's follow the given steps to deploy the app using the Appium app:

1. Launch the Appium GUI app.
2. Click on the apple icon and it will open the **iOS Settings** popup.

3. Enter the details:
   - **App Path** - path to the test app which we generated earlier.
   - Select **Force Device** and choose iPhone 6 from the dropdown.
   - Select **Platform Version** to be 8.4:

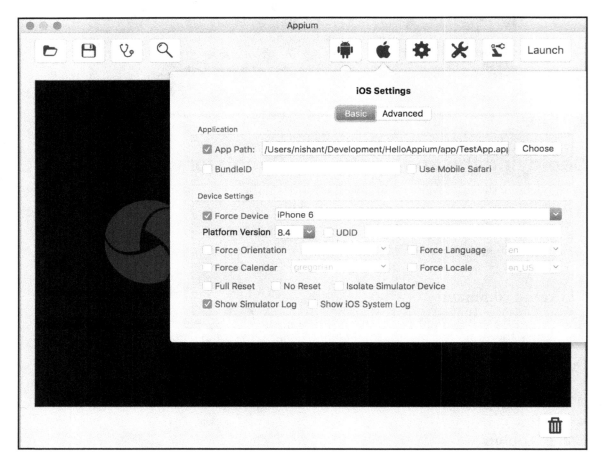

4. Once done, click on the apple icon to close the **iOS Settings** screen.
5. Click on **Launch** on the Appium GUI.

6. This will start the Appium server with the following logs:

```
Launching Appium with command: '/Applications/Appium 2.app/Contents/Resources/node/bin/node' appium/build/
lib/main.js --session-override --pre-launch --debug-log-spacing --strict-caps --platform-version "8.4" --
platform-name "iOS" --app "/Users/nishant/Development/HelloAppium/app/TestApp.app" --device-name "iPhone 6"

[Appium] Welcome to Appium v1.5.3

[Appium] Non-default server args:

[Appium]    sessionOverride: true
[Appium]    launch: true
[Appium]    enforceStrictCaps: true
[Appium]    debugLogSpacing: true
[Appium]    platformName: 'iOS'
[Appium]    platformVersion: '8.4'
[Appium]    deviceName: 'iPhone 6'
[Appium]    app: '/Users/nishant/Development/HelloAppium/app/TestApp.app'
[Appium] Deprecated server args:
[Appium]    --platform-name => --default-capabilities '{"platformName":"iOS"}'
[Appium]    --platform-version => --default-capabilities '{"platformVersion":"8.4"}'
[Appium]    --device-name => --default-capabilities '{"deviceName":"iPhone 6"}'
[Appium]    --app => --default-capabilities '{"app":"/Users/nishant/Development/HelloAppium/app/TestApp.app"}'
[Appium] Default capabilities, which will be added to each request unless overridden by desired capabilities

[Appium]    platformName: 'iOS'

[Appium]    platformVersion: '8.4'
[Appium]    deviceName: 'iPhone 6'
[Appium]    app: '/Users/nishant/Development/HelloAppium/app/TestApp.app'

[Appium] Appium REST http interface listener started on 0.0.0.0:4723
```

7. Click on the inspector icon 🔍 .

8. It will launch the Appium inspector with the app screen captured:

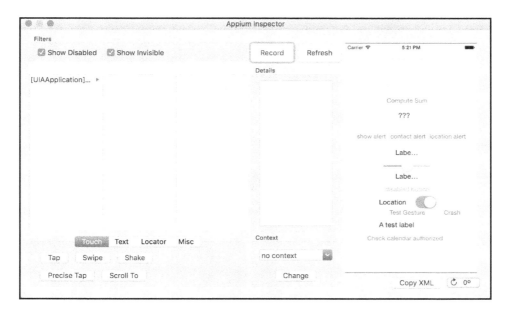

So this step deploys the app on the iOS simulator and launches the app. We can click on any of the UI elements on the right of the preceding screen and the left panel will be loaded with the respective attribute details.

Generating Boilerplate code for iOS

Once we have the preceding Appium set up does and it runs successfully to deploy the app on the iOS simulator, we can use the boiler plate code to implement the first step which is about launching the app. The precedingly written scenario in case of android is contextual to the app; let's write a unique scenario for the iOS app to understand the concepts under a new feature file and name it `Sample_ios.feature`:

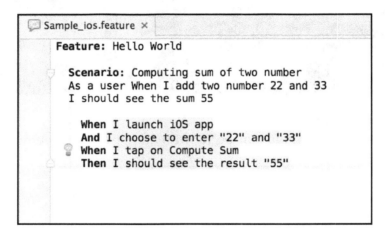

Following is the snippet which you can copy after creating a new feature file for iOS features:

```
Feature: Hello World

Scenario: Computing sum of two number
 As a user When I add two number 22 and 33
 I should see the sum 55

When I launch iOS app
And I choose to enter "22" and "33"
When I tap on Compute Sum
Then I should see the result "55"
```

So the preceding scenario is performing in following steps:

1. Launch the app.
2. Enter two numbers *22* and *33* in the specified text-box.
3. Tap on Compute Sum.
4. Verify result as *55*.

Let's implement the first step which is launching the iOS app. Remember we have the Appium session running and the app is already launched on the simulator. Click on the **Record** button on the Appium inspector screen. This will generate the boiler plate code as shown:

```java
public class {scriptName} {
    public static void main(String[] args) {
        DesiredCapabilities capabilities = new DesiredCapabilities();
        capabilities.setCapability("appium-version", "1.0");
        capabilities.setCapability("platformName", "iOS");
        capabilities.setCapability("platformVersion", "8.4");
        capabilities.setCapability("deviceName", "iPhone 6");
        capabilities.setCapability("app", "/Users/nishant/Development/HelloAppium/app/TestApp.app");
        wd = new AppiumDriver(new URL("http://0.0.0.0:4723/wd/hub"), capabilities);
        wd.manage().timeouts().implicitlyWait(60, TimeUnit.SECONDS);
        wd.close();
    }
}
```

So if we notice the important desired capabilities, they are:

1. `platformName` - iOS
2. `platformVersion` - 8.4
3. `deviceName` - iPhone 6
4. `app` - Actual app path

Let's implement the first step `When I launch iOS app`; we need to copy the preceding code generated as boiler plate and make some minor tweaks as shown:

```
@When("^I launch iOS app$")
public void iLaunchIOSApp() throws Throwable {
  DesiredCapabilities capabilities = new DesiredCapabilities();
  capabilities.setCapability("appium-version", "1.0");
  capabilities.setCapability("platformName", "iOS");
  capabilities.setCapability("platformVersion", "8.4");
  capabilities.setCapability("deviceName", "iPhone 6");
  capabilities.setCapability("app",
  "/Users/nishant/Development/HelloAppium/app/TestApp.app");
  AppiumDriver wd = new AppiumDriver(new
  URL("http://0.0.0.0:4723/wd/hub"), capabilities);
  wd.manage().timeouts().implicitlyWait(60, TimeUnit.SECONDS);
}
```

This gives us the driver instance `wd` which can be used for all the test purpose. Let's learn how to run the preceding steps.

So, the steps to run the feature file are these:

1. Start the terminal and type in `appium --session-override`.
2. Navigate to IntelliJ and open the feature file `Sample_ios.feature`.
3. Right click on the feature file and choose the run feature option:

> Run 'Feature: Sample_ios'   ^⇧F10

This will launch the iOS simulator and run the test which is basically going to deploy the app and launch the app. The test would fail because other steps are yet to be implemented.

We need to perform the above demonstrated refactoring to extract the AppiumDriver instance and make it a class variable. Once it is done, let's automate other steps to complete the scenario automation. Next step to get automated is :

```
And I choose to enter "22" and "33"
```

We need to figure out the locators of the text box. Launch the appium GUI app, previously made iOS settings would persist. Click on **Start**, this will start the appium server. We need to wait till the appium server is started and the console log is shown as below:

And then click on the **Appium Inspector** icon which would start the iOS simulator and deploy the app on the same. This would deploy the app, and launch the appium inspector window. Click on the first text box field on the right of the appium inspector window, it will load the UI hierarchy and the attributes in the left pane, as shown in image below:

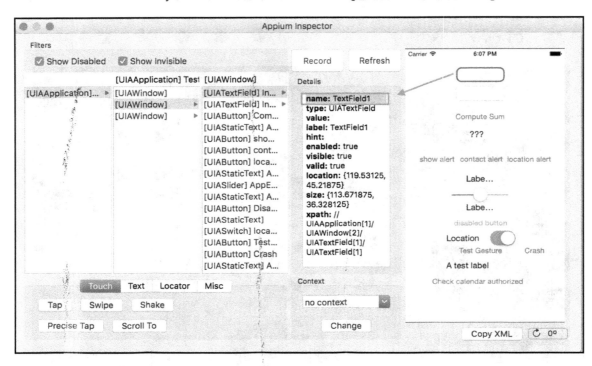

We can use the property name for identifying the first text box by using the method findElementByAccessibilityId() and pass the property name as the parameter.

Step implementation for the above mentioned step would be:

```
@And("^I choose to enter \"([^\"]*)\" and \"([^\"]*)\"$")
public void iChooseToEnterAnd(String num1, String num2) throws Throwable {
wd.findElementByAccessibilityId("TextField1").sendKeys(num1);
wd.findElementByAccessibilityId("TextField2").sendKeys(num2);
}
```

In the above code, we are trying to find a field using it's accessibility identifier and then pass a value in that field. Similarly we can implement the other two steps using accessibility identifier :

```
When I tap on Compute Sum
Then I should see the result "55"
```

Below is how the iOSPageSteps.Java file would look like after the implementation:

```
public class iOSPageSteps {

private AppiumDriver appiumDriver;

@When("^I launch iOS app$")
public void iLaunchIOSApp() throws Throwable {
 DesiredCapabilities capabilities = new DesiredCapabilities();
 capabilities.setCapability("appium-version", "1.0");
 capabilities.setCapability("platformName", "iOS");
 capabilities.setCapability("platformVersion", "8.4");
 capabilities.setCapability("deviceName", "iPhone 6");
 capabilities.setCapability("app",
 "/Users/nishant/Development/HelloAppium/app/TestApp.app");
 appiumDriver = new AppiumDriver(new URL
 ("http://0.0.0.0:4723/wd/hub"), capabilities);
 appiumDriver.manage().timeouts()
 .implicitlyWait(60, TimeUnit.SECONDS);
 }

@And("^I choose to enter \"([^\"]*)\" and \"([^\"]*)\"$")
public void iChooseToEnterAnd(String num1, String num2) throws Throwable {
appiumDriver.findElementByAccessibilityId("TextField1").sendKeys(num1);
appiumDriver.findElementByAccessibilityId("TextField2").sendKeys(num2);
 }

@When("^I tap on Compute Sum$")
public void iTapOnComputeSum() throws Throwable {
appiumDriver.findElementByAccessibilityId("ComputeSumButton").click();
 }

@Then("^I should see the result \"([^\"]*)\"$")
public void iShouldSeeTheResult(String result) throws Throwable {
 Assert.assertEquals(result,
 appiumDriver.findElementByAccessibilityId("Answer").getText());
 }

}
```

This completes the automation of the scenario for iOS app. In the upcoming chapters, we can perform the general refactoring on the above generated code for iOS app. However, some of the code which are android specific will not work for iOS app.

Summary

In this chapter, we introduced you to basic a Appium Java project and how to use Cucumber to write the test. Also, we briefly discussed the importance of Cucumber and how it helps capture the system's behavior. We added Appium and Cucumber dependency in `Gradle` file. We were also introduced to the desired capabilities class, which tells the Appium server what session we are interested in. We saw how the desired capabilities differ from hybrid app and mobile web app. We also got to know about `browserName` and the values it can take.

In the next chapter, we will take a detailed look into the `Desired Capabilities` class and how to vary the parameters to suit our testing needs. Also, we will refactor the test to start the server programmatically and see the arguments it can take.

4
Understanding Desired Capabilities

In the last chapter, we saw that the boilerplate code generated by Appium Inspector had a bunch of lines that used the `DesiredCapabilities` class and passed a certain set of keys and values to the Appium server. In this chapter, we will take a detailed look at the following:

- Appium server arguments
- Desired capabilities for Android
- Desired capabilities for iOS
- iOS XCUITest related iOS capabilities

Before we take a dive in there, we will refactor the code written in the last chapter and introduce the concept of hooks before and *after*, which acts like setup and tear down and will take care of starting the Appium server programmatically and then stopping it.

In test automation with Appium, all the commands are executed in the context of a **session**. A session is initiated by a client with a server in ways either specific to Android or iOS and with a JSON object called desired capabilities.

Let's refactor the existing code to add handling the Appium server through code.

Refactoring -1

Note: We will take one of the feature files or code bases (Android in this case) to demonstrate some of the concepts while refactoring. This can also be followed with the other iOS code written.

In this first refactoring, we will remove the manual dependency of starting and stopping the Appium server, and we will do it programmatically:

- Create a new class called `StartingSteps` under the `steps` package
- In the `StartingSteps` class, create two empty methods, called `startAppiumServer` and `stopAppiumServer`:

```
public void startAppiumServer(){
//code to start appium server
}

public void stopAppiumServer(){
// Code to stop appium server
}
```

At this point in time, we need to know the concept of hooks in cucumber. So basically, cucumber gives you a number of hooks, which allow one to run certain code at a certain point in the test life cycle. These hooks can be used and defined in a class file in the steps folder. However, cucumber doesn't mandate the location. So, the two hooks that we will use are these:

- **Before**: Before hooks will run before the execution of each scenario. They execute before the first step mentioned in each scenario; hence, they can potentially act as a common setup for all the tests. We can have multiple before hooks, and they will run in the same order as they are declared.
- **After**: After hooks run after the last step of each scenario. they run irrespective of the outcome of the last step, whether the last step is a success or failure.

Let's put the @Before tag for the startAppiumServer method and the @After tag for stopAppiumServer method. While resolving, ensure that you use the cucumber.api.java, highlighted here:

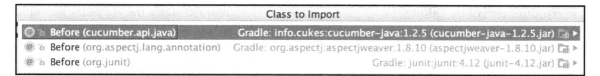

- The next step is to add code that will start the Appium server. Appium exposes AppiumDriverLocalService, which will basically let you start and stop the server.

- Add the following code to the startAppiumServer method. So, this builds the default Appium service, which means that the **IP address** will remain **0.0.0.0** and the **port** will be **4723**:

```
@Before
public void startAppiumServer() throws IOException {
    AppiumDriverLocalService appiumService =
    AppiumDriverLocalService.buildDefaultService();
    appiumService.start();
}
```

- To implement the stopAppiumServer method, we need appiumService to be in instance variable; so, highlight the first row above and select **Refactor** > **Extract** > **Field**.

- Once done in the stopAppiumServer method, implement the following code. This stops the Appium server:

```
@After
public void stopAppiumServer() {
    appiumService.stop();
}
```

- Once done, stop any instance of Appium server running via the terminal (Command prompt in case of Windows) or Appium GUI app.
- Navigate to the feature file, **right-click**, and select the **Run 'Feature:Sample'** option.

You should be able to run both the scenarios without having to start the Appium server manually.

Server argument

Desired capabilities are sent by the client to the server via JSON objects by requesting the automation session we intend to have. Now, with the preceding code refactor, we can start the Appium server by calling the `start()` method. This method starts the Appium server assuming the `node` and `appium.js` files to be in certain locations and using port **4723**. If you have tweaked `node.js` installation, your Appium install path is not the same, or if you want to start Appium server on different ports, we can use `AppiumServiceBuilder` to override these inputs:

```
appiumService = AppiumDriverLocalService.buildService(new
AppiumServiceBuilder()
        .usingDriverExecutable(new File(("/path/to/node")))
        .withAppiumJS(new File(("/path/to/appium")))
        .withIPAddress("127.0.0.1")
        .usingPort(port)
        .withArgument(argument)
        .withLogFile(new File("path/to/log/file")));
appiumService.start();
```

`AppiumServiceBuilder` also gives you other options to override, which are as shown here:

`withAppiumJS(File appiumJS)`	AppiumServiceBuilder
`usingAnyFreePort()`	AppiumServiceBuilder
`usingDriverExecutable(File nodeJSExecutabl…`	AppiumServiceBuilder
`usingPort(int port)`	AppiumServiceBuilder
`withArgument(ServerArgument argument)`	AppiumServiceBuilder
`withArgument(ServerArgument argument, Stri…`	AppiumServiceBuilder
`withCapabilities(DesiredCapabilities capab…`	AppiumServiceBuilder
`withEnvironment(Map<String, String> enviro…`	AppiumServiceBuilder
`withIPAddress(String ipAddress)`	AppiumServiceBuilder
`withLogFile(File logFile)`	AppiumServiceBuilder
`withStartUpTimeOut(long time, TimeUnit tim…`	AppiumServiceBuilder

If you note the `withArgument(ServerArgument argument)` method, Appium server can take input from the following list of arguments. We will mention some important ones that one should be aware of, though Appium lists a bunch of them that can be found on their site at `https://github.com/appium/appium/blob/master/docs/en/writing-running-ap pium/server-args.md`(The following descriptions/definitions of certain flags are cross referenced from Appium website):

Flag	Description
`--address`	This is the IP address to listen on. Default value: `0.0.0.0` (usage example, `appium --address 192.168.1.1`)
`--port`	This is the port to listen on. Default value: `4723` (usage example, `appium --port 4726`)
`--session-override`	This enables session override. Default value: `False` (usage example, `appium --session-override`)
`--log`	This is to send log output to the file specified. Default value: `null` (usage example, `appium --log path/to/appium.log`)
`--selendroid-port`	This is the port used for communication with Selendroid. Default value: `8080` (usage example, `appium --selendroid-port 8282`)
`--bootstrap-port`	This is the port to use on devices to talk to Appium. This is an Android-only. Default value: `4724` (usage example, `appium --bootstrap-port 4728`)
`--webhook`	This is to send log output to the HTTP listener specified. Default value: `null` (usage example, `appium --webhook localhost:9876`)

So, let's do another refactoring where we replace the `AppiumDriverLocalService.buildDefaultService()` method to build our own Appium service.

Refactoring -2

Let's refactor the given line and change the `buildDefaultService()`; instead, we will use the `buildService()` method, which will take some of the server flags discussed earlier:

```
AppiumDriverLocalService appiumService =
AppiumDriverLocalService.buildDefaultService();
```

Ensure that the path to node and Appium is what you found on your machine. To find the path on your Mac machine, launch Terminal and run the `which appium` and `which node` commands to get the path:

```
Last login: Tue Jan 24 10:35:56 on ttys000
[➔ ~ which appium
/usr/local/bin/appium
[➔ ~ which node
/usr/local/bin/node
➔ ~ ▮
```

Also, the path to the log file can be relative to the project folder. So, now the method should look as shown in the following code:

```
@Before
public void startAppiumServer() throws IOException {

    int port = 4723;

    appiumService = AppiumDriverLocalService.buildService(new
    AppiumServiceBuilder()
            .usingDriverExecutable(new File(("/usr/local/bin/node")))
            .withAppiumJS(new File(("/usr/local/bin/appium")))
            .withIPAddress("0.0.0.0")
            .usingPort(port)
            .withArgument(GeneralServerFlag.SESSION_OVERRIDE)
            .withLogFile(new File("build/appium.log")));
    appiumService.start();
}
```

For a Windows user, we need to get the path of executables of Node JS and Appium. So, instead of having two methods in the same class, we will add a check for the OS type, and we will start the Appium server based on the OS type. So, the preceding code will look as illustrated; dependency is `nodeJS_Path`, and `appiumJS_Path` should be defined based on your machine:

```
@Before
public void startAppiumServer() throws IOException {

    int port = 4723;
    String nodeJS_Path = "C:/Program Files/NodeJS/node.exe";
    String appiumJS_Path = "C:/Program
    Files/Appium/node_modules/appium/bin/appium.js";

    String osName = System.getProperty("os.name");

    if (osName.contains("Mac")) {
        appiumService = AppiumDriverLocalService.buildService(new
        AppiumServiceBuilder()
                .usingDriverExecutable(new
                File(("/usr/local/bin/node")))
                .withAppiumJS(new File(("/usr/local/bin/appium")))
                .withIPAddress("0.0.0.0")
                .usingPort(port)
                .withArgument(GeneralServerFlag.SESSION_OVERRIDE)
                .withLogFile(new File("build/appium.log")));
    } else if (osName.contains("Windows")) {
        appiumService = AppiumDriverLocalService.buildService(new
        AppiumServiceBuilder()
                .usingDriverExecutable(new File(nodeJS_Path))
                .withAppiumJS(new File(appiumJS_Path))
                .withIPAddress("0.0.0.0")
                .usingPort(port)
                .withArgument(GeneralServerFlag.SESSION_OVERRIDE)
                .withLogFile(new File("build/appium.log")));
    }
    appiumService.start();
}
```

In the preceding code, we noticed that the SESSION_OVERRIDE belongs to GeneralServerFlag; apart from that, there are a lot of other flags as well, which can be chosen from this list:

SESSION_OVERRIDE	GeneralServerFla
ASYNC_TRACE	GeneralServerFlag
CALLBACK_ADDRESS	GeneralServerFlag
CALLBACK_PORT	GeneralServerFlag
CONFIGURATION_FILE	GeneralServerFlag
DEBUG_LOG_SPACING	GeneralServerFlag
LOCAL_TIMEZONE	GeneralServerFlag
LOG_LEVEL	GeneralServerFlag
LOG_NO_COLORS	GeneralServerFlag
LOG_TIMESTAMP	GeneralServerFlag
NO_PERMS_CHECKS	GeneralServerFlag
PRE_LAUNCH	GeneralServerFlag
ROBOT_ADDRESS	GeneralServerFlag
ROBOT_PORT	GeneralServerFlag
SHELL	GeneralServerFlag
SHOW_CONFIG	GeneralServerFlag
STRICT_CAPS	GeneralServerFlag
TEMP_DIRECTORY	GeneralServerFlag
valueOf(String name)	GeneralServerFlag
WEB_HOOK	GeneralServerFlag
values()	GeneralServerFlag[]

For example, we can choose to pass the ROBOT_ADDRESS flag and pass the device ID or UDID to run the test on a certain device connected to the machine:

```
.withArgument(GeneralServerFlag.ROBOT_ADDRESS, udid)
```

We will use the udid feature when we need to execute the test on physical devices or more than one device with similar configurations. We will read about this in detail in the upcoming chapter. Let's move on to the server capabilities.

Server capabilities

Testing will always be performed in a fixed context with respect to Appium server and that context will be set by desired capabilities. There are some mandatory desired capabilities and some are device OS-specific, such as Android or iOS. If you navigate to the `HomePageSteps` class file and the `iLaunchQuikr()` method, you will note the following mentioned lines:

```
capabilities.setCapability("platformName", "Android");
capabilities.setCapability("platformVersion", "5.1");
capabilities.setCapability("deviceName", "Nexus");
```

So, `platformName`, `platformVersion`, and `deviceName` are mandatory desired capabilities, which form the basis of mobile automation.

Here are some of the capabilities that are applicable for both iOS as well as Android devices:

Capability	Description & Usage
`automationName`	This is used to specify the automation engine to be used. **Values**: `Appium` (default) or `Selendroid`
`platformName`	This is used to specify which mobile OS platform to use for the session. `capabilities.setCapability("platformName", "Android");`
`platformVersion`	This is used to specify which mobile OS version to be used for the session, and it should match the emulator or device under test. `capabilities.setCapability("platformVersion", "5.1");`
`deviceName`	This is used to specify which mobile device or emulator is in use. For Android, any string can be passed. `capabilities.setCapability("deviceName", "Nexus");`
`app`	This is used to specify the absolute local path to an `.ipa` or `.apk` file. `capabilities.setCapability("app", "/Users/Development/HelloAppium/app.apk");`

`browserName`	This is used to specify the name of the mobile web browser to automate. **Values:** `Safari`, `Chrome`, `Chromium,` and `Browser` `capabilities.setCapability("browserName", "Browser");`
`newCommandTimeout`	This is used to specify the time (in seconds) Appium will wait for a new command from the client before assuming that the client quit and ending the session. `capabilities.setCapability("newCommandTimeout", 120);`
`language`	This is used to specify the language for the simulator or emulator. (usage example, `fr`)
`locale`	This is used to specify the locale to set for the iOS Simulator For example, `fr_CA`
`udid`	This is used to specify the unique device identifier of the connected physical device For example, `1ae203187fc012g` `capabilities.setCapability("udid", 1ae203187fc012g);`
`orientation`	This is used to specify the orientation for simulator or emulator. `LANDSCAPE` or `PORTRAIT`
`autoWebview`	This is used to move directly into Webview context; **default value is false**. **Values:** `true` and `false` `capabilities.setCapability("autoWebview", true);`
`noReset`	Don't reset the app state; the **default value is false**. This flag is used when you don't want to reset the app state. **Values:** true and false `capabilities.setCapability("noReset", true);`
`fullReset`	Reset app state; in iOS, delete the entire simulator folder. In Android, reset the app state by uninstalling the app and clearing all data; full reset requires an app capability. The **default value is false**. **Values:** `true` and `false` `capabilities.setCapability("fullReset", true);`

Refactoring -3

Let's add a couple of desired capabilities to our test now. One of the important concepts of test automation is to start with a clean slate; hence, we will add `fullReset` capabilities. Also, the Appium server waits for 60 secs for the new command by default, and then it quits the session because of inactivity; so we will tweak it a little to wait for 120 seconds now:

- `fullReset`: This is to reset the app state between each test
- `newCommandTimeout`: This is to stop the session from quitting if new commands are not passed within 120 seconds

Let's add these capabilities in the `HomePageSteps` class file under the `iLaunchQuikrApp()` method:

```
@When("^I launch Quikr app$")
public void iLaunchQuikrApp() throws Throwable {
    DesiredCapabilities capabilities = new DesiredCapabilities();
    capabilities.setCapability("platformName", "Android");
    capabilities.setCapability("platformVersion", "5.1");
    capabilities.setCapability("deviceName", "Nexus");
    capabilities.setCapability("fullReset", true);
    capabilities.setCapability("newCommandTimeout", 120);
    capabilities.setCapability("app",
    "/Users/nishant/Development/HelloAppium/app/quikr.apk");
    appiumDriver = new AppiumDriver(new
    URL("http://0.0.0.0:4723/wd/hub"), capabilities);
    appiumDriver.manage().timeouts().implicitlyWait(60,
    TimeUnit.SECONDS);
}
```

We can try running the scenario again, and it will work seamlessly. After every refactoring, ensure that you execute the code so that you know the impact of the change.

Android-only capabilities

When we are performing Android app automation, there are a bunch of Android-specific capabilities that can be used to set the session. Here's a complete list of **Android-only** capabilities:

Capability	Description and values
appActivity	Used to specify the activity name for the Android activity you want to launch from your package. `capabilities.setCapability("appActivity", MainActivity);`
appPackage	Used to specify the Java package name of the Android app you want to run. `capabilities.setCapability("appPackage", com.example.android.mySampleApp);`
appWaitActivity	Used to specify the activity name for the Android activity you want to wait for. `capabilities.setCapability("appWaitActivity", SplashActivity);`
appWaitPackage	Used to specify the Java package of the Android app you want to wait for. `capabilities.setCapability("appWaitPackage", com.example.android.myApp);`
appWaitDuration	Used to specify timeout in milliseconds used to wait for the appWaitActivity to launch (default is 20000). `capabilities.setCapability("appWaitDuration", 30000);`
deviceReadyTimeout	Used to specify timeout in seconds while waiting for the device to become ready. `capabilities.setCapability("deviceReadyTimeout", 5);`
androidCoverage	Used to specify a fully qualified instrumentation class. Passed to -w in adb shell am instrument -e coverage true -w. `com.my.Pkg.instrumentation.MyInstrumentation`
enablePerformanceLogging	Used to enable ChromeDriver's performance logging only for Chrome and webview (default `false`). `capabilities.setCapability("enablePerformanceLogging", true);`
androidDeviceReadyTimeout	Used to specify timeout in seconds used to wait for a device to become ready after booting. `capabilities.setCapability("androidDeviceReadyTimeout", 30);`

androidInstallTimeout	Used to specify timeout in milliseconds, it is used to wait for an apk to install to the device. Defaults to 90000. `capabilities.setCapability("androidInstallTimeout", 60000);`
adbPort	Used to specify the port used to connect to the ADB server (default 5037).
androidDeviceSocket	Used to specify the Devtools socket name. It is needed only when the tested app is a Chromium-embedding browser. The socket is opened by the browser, and ChromeDriver connects to it as a client, for example, `chrome_devtools_remote`.
avd	Used to specify the name of the avd to launch. `capabilities.setCapability("avd", "Nexus6");`
avdLaunchTimeout	Used to specify how long to wait in milliseconds for an avd to launch and connect to ADB (default 120000). `capabilities.setCapability("avdLaunchTimeout", 300000);`
avdReadyTimeout	Used to specify how long to wait in milliseconds for an avd to finish its boot animations (default 120000). `capabilities.setCapability("avdReadyTimeout", 300000);`
avdArgs	Used to specify additional emulator arguments, it is used when launching an avd. Usage example, `-netfast`
useKeystore	Used to specify a custom keystore to sign apks; default false. `capabilities.setCapability("useKeystore", true);`
keystorePath	Used to specify the path to custom keystore; default . Usage example, `e`
keystorePassword	Used to specify the password for custom keystore. For example, `foo`
keyAlias	Used to specify the alias for key.
keyPassword	Used to specify the password for key. For example, `foo`
chromedriverExecutable	Used to specify the absolute local path to webdriver executable (if Chromium embedder provides its own webdriver, it should be used instead of the original ChromeDriver bundled with Appium). Usage example, `r`
autoWebviewTimeout	Used to specify the amount of time to wait for Webview context to become active, in `ms`; defaults to 2000.
intentAction	Used to specify the intent action that will be used to start activity (default `android.intent.action.MAIN`). For e, `android.intent.action.MAIN` and `android.intent.action.VIEW`

`intentCategory`	Used to specify intent category which will be used to start activity (default `android.intent.category.LAUNCHER`). For e, `android.intent.category.LAUNCHER` and `android.intent.category.APP_CONTACTS`
`intentFlags`	Used to specify flags that will be used to start activity (default `0x10200000`). For example, `0x10200000`
`optionalIntentArguments`	Used to specify additional intent arguments that will be used to start activity; refer to Intent arguments. For example, `--esn`, `--ez`, and more.
`unicodeKeyboard`	Used to enable Unicode input, default is `false`.
`resetKeyboard`	Used to reset the keyboard to its original state after running Unicode tests with the d capability. It is ignored if used alone; default is `false`.
`dontStopAppOnReset`	This doesn't stop the process of the app under test before starting the app using adb. If the app under test is created by another anchor app, setting this to false allows the process of the anchor app to be still alive during the start of the test app using adb. In other words, with t set to true, we will not include the S flag in the adb shell start call. With this capability omitted or set to false, we include the `-S` flag; default is `false`.
`noSign`	Used to skip checking and signing of the app with debug keys, it will work only with r and not with selendroid; defaults to `false`.
`ignoreUnimportantViews`	Used to call the) UiAutomator function. This capability can speed up test execution since accessibility commands will run faster, ignoring some elements. The ignored elements will not be findable, which is why this capability has also been implemented as a toggle-able setting as well as a capability; defaults to `false`.
`disableAndroidWatchers`	Used to disable Android watchers that watch for applications not responding and application crash; this will reduce CPU usage on Android device/emulator. This capability will work only with UiAutomator and not with Selendroid; default is `false`.
`chromeOptions`	Used to allow passing the s capability for ChromeDriver. For more information, refer to ChromeOptions at: `https://sites.google.com/a/chromium.org/chromedriver/capabilities`
`recreateChromeDriverSessions`	Used to kill a ChromeDriver session when moving to a non-ChromeDriver webview; defaults to `false`.
`nativeWebScreenshot`	Used to specify web context and use the native (adb) method for taking a screenshot; defaults to `false`.
`androidScreenshotPath`	Used to specify the name of the directory on the device in which the screenshot will be put; defaults to `/data/local/tmp`. For example, `/sdcard/screenshots/`
`autoGrantPermissions`	Used to have Appium automatically determine which permissions your app requires and grant them to the app on install; defaults to `false`.

Let's refactor the code to use some of the preceding capabilities.

Refactoring -4

When we want to use the app that is already installed on the emulator, we need to use the `appPackage` and `appActivity` capabilities and not use the `app` capability. Let's look at the code we have written till now for starting an app in the `HomePageSteps`.

```
@When("^I launch Quikr app$")
    public void iLaunchQuikrApp() throws Throwable {
        DesiredCapabilities capabilities = new DesiredCapabilities();
        capabilities.setCapability("platformName", "Android");
        capabilities.setCapability("platformVersion", "5.1");
        capabilities.setCapability("deviceName", "Nexus");
        capabilities.setCapability("noReset", false);
        capabilities.setCapability("fullReset", true);
        capabilities.setCapability("app",
        "/Users/nishant/Development/HelloAppium/app/quikr.apk");

        appiumDriver = new AppiumDriver(new
        URL("http://0.0.0.0:4723/wd/hub"), capabilities);
        appiumDriver.manage().timeouts().implicitlyWait(60,
        TimeUnit.SECONDS);
    }
```

When we execute the preceding code, it performs a full reset and deploys the app every time. Let's modify this code to use the installed app and perform a fast reset of the app. Comment or delete the lines, as shown:

```
        // capabilities.setCapability("noReset", false);
        // capabilities.setCapability("fullReset", true);
        // capabilities.setCapability("app",
        "/Users/nishant/Development/HelloAppium/app/quikr.apk");
```

Also, add the following lines. Once done, run the test by right-clicking on the feature file and selecting the **Run Scenario...** option:

```
capabilities.setCapability("appPackage", "com.quikr");
capabilities.setCapability("appActivity", "com.quikr.old.SplashActivity");
```

If we have an Android virtual device created (using Android SDK, we created one in Chapter 2, *Setting Up the Machine*), we can use the following code that will start the emulator first and then run the test:

```
capabilities.setCapability("avd", "Nexus6_API_24");
capabilities.setCapability("avdReadyTimeout", 180000);
```

So, with the preceding change, the iLaunchQuikrApp method will look like this:

```
@When("^I launch Quikr app$")
public void iLaunchQuikrApp() throws Throwable {
    DesiredCapabilities capabilities = new DesiredCapabilities();
    capabilities.setCapability("platformName", "Android");
    capabilities.setCapability("platformVersion", "5.1");
    capabilities.setCapability("deviceName", "Nexus");
    capabilities.setCapability("newCommandTimeout", 120);

    // Launches the below android virtual device and waits for 120
    seconds for AVD to be ready
    capabilities.setCapability("avd", "Nexus6_API_24");
    capabilities.setCapability("avdReadyTimeout", 120000);

    capabilities.setCapability("appPackage", "com.quikr");
    capabilities.setCapability("appActivity",
    "com.quikr.old.SplashActivity");

    appiumDriver = new AppiumDriver(new
    URL("http://0.0.0.0:4723/wd/hub"), capabilities);
    appiumDriver.manage().timeouts().implicitlyWait(60,
    TimeUnit.SECONDS);
}
```

Once we make the preceding changes, we should be able to run it; this will launch the Android virtual device as well before triggering the test on it.

Let's take a look at the iOS only capabilities. Even though the app behavior is more or less similar across the devices, Appium exposes a bunch of different sets of capabilities for iOS.

iOS-only capabilities

When we are performing iOS app automation, there are a bunch of iOS specific capabilities that can be used to set the session. Here's a complete list of iOS-only capabilities. Recently, Appium implemented support for **XCUITest**; hence, there are bunch of other capabilities that are **XCUITest** specific:

Capability	Description and values
`calendarFormat`	Used to specify the calendar format to set for the iOS Simulator. For example, `gregorian`
`bundleId`	Used to specify the bundle ID of the app under test. For example, `io.HelloiOS.TestApp`
`udid`	Used to specify the unique device identifier of the connected device. For example, `1ae203187fc012g`
`launchTimeout`	Used to specify the amount of time (in ms) to wait for instruments. For example, `5000`
`locationServicesEnabled`	Used to force location services to be either on or off. The default behavior is to keep the current simulator setting. Value: `true` or `false`
`locationServicesAuthorized`	Used to set location services to be authorized or not authorized for an app via plist so that the location services alert doesn't pop up. The default is to keep the current simulator setting. Value: `true` or `false`
`autoAcceptAlerts`	Used to accept all iOS alerts automatically if they pop up. This includes privacy access permission alerts (For example, location, contacts, and photos); default is `false`. It does not work on XCUITest-based tests. Value: `true` or `false`

`autoDismissAlerts`	Used to dismiss all iOS alerts automatically if they pop up. This includes privacy access permission alerts (For example, location, contacts, and photos); default is `false`. It doesn't work on XCUITest-based tests. Value: true or false
`nativeInstrumentsLib`	Used to use native instruments b (that is, disable instruments-without-delay). Value: `true` or `false`
`nativeWebTap`	Used to enable "real", non-javascript-based web taps in Safari; default is `false`. Value: true or false
`safariInitialUrl`	Used to specify initial Safari URL; default is a local welcome page and its simulator-only capability, and works on version 8.1 onward. For example, m
`safariAllowPopups`	Used to allow JavaScript to open new windows in Safari. By default, it keeps the current sim setting. Value: `true` or `false`
`safariIgnoreFraudWarning`	Used to prevent Safari from showing a fraudulent website warning. **By default, it keeps the current sim setting.** It's a simulator-only feature. Value: `true` or `false`
`safariOpenLinksInBackground`	Used to specify whether Safari should allow links to open in new windows. **By default, it keeps the current sim setting.** It's a simulator-only feature. Value: `true` or `false`
`keepKeyChains`	Used to specify whether to keep keychains (Library/Keychains) when an Appium session is started/finished. It's a simulator-only feature. Value: `true` or `false`
`localizableStringsDir`	Used to specify where to look for localizable strings; `default is en.lproj`. `en.lproj`

processArguments	Used to specify arguments to pass to the AUT using instruments. For example, `-myflag`
interKeyDelay	Used to specify the delay, in ms, between keystrokes sent to an element when typing. For example, `100`
showIOSLog	Used to specify whether to show any logs captured from a device in the Appium logs; default is `false`. Value: `true` or `false`
sendKeyStrategy	Used to specify the strategy to use to type text into a t field; simulator default: `oneByOne`, real device default: `grouped`. Value: `e`, `grouped`, or `setValue`
screenshotWaitTimeout	Used to specify maximum timeout in seconds to wait for a screenshot to be generated; default is `10`. For example, `5`
waitForAppScript	Used to specify the iOS automation script used to determine whether the app has been launched; by default, the system waits for the page source not to be empty. The result must be a boolean. For example, `true;`, `target.elements().length > 0;`, `$.delay(5000); true;`
webviewConnectRetries	Used to specify the number of times we are to send a connection message to the remote debugger to get a webview; default: `8`. For example, `3`
appName	Used to specify the display name of the application under test. It is used to automate backgrounding the app in iOS 9+. For example, `UICatalog`

customSSLCert	Used to specify an SSL certificate to simulator. It's a simulator-only feature. For example, -----BEGIN CERTIFICATE-----QWEQWEQWEKg... -----END CERTIFICATE-----

Here are the iOS XCUITest related iOS capabilities:

Capability	Description	Values
processArguments	Process arguments and environment that will be sent to the t server.	{ args: ["a", "b", "c"] , env: { "a": "b", "c": "d" } } or '{"args": ["a", "b", "c"], "env": { "a": "b", "c": "d" }}'
wdaLocalPort	This value, if specified, will be used to forward traffic from the Mac host to real iOS devices over a USB. The default value is the same as the port number used by WDA on device.	For example, 8100
showXcodeLog	To display the output of the Xcode command used to run the tests. If this is true, there will be lots of extra logging at startup; it defaults to false.	For example, true
iosInstallPause	Time in milliseconds to pause between installing the application and starting WebDriverAgent on the device. It is used particularly for larger applications, and defaults to 0.	For example, 8000

xcodeConfigFile	Full path to an optional Xcode configuration file that specifies the code signing identity and team for running the WebDriverAgent on the real device.	For example, g
keychainPath	Full path to the private development key exported from the system keychain. It is used in conjunction with d when testing on real devices.	For example, 2
keychainPassword	Password for unlocking the keychain specified in h.	For example, super awesome password
scaleFactor	Simulator scale factor. This is useful to have if the default resolution of the simulated device is greater than the actual display resolution. So you can scale the simulator to see the whole device screen without scrolling.	Acceptable values are ' ', '0.75', '0.5', '0.33', and '0.25'; the value should be a string
preventWDAAttachments	Sets read-only permissons to the s subfolder of the WebDriverAgent root inside Xcode's DerivedData. This is necessary to prevent the XCTest framework from creating tons of unnecessary screenshots and logs, which are impossible to shut down using programming interfaces provided by Apple.	Setting the capability to true will set Posix permissions of the folder to 555, and false will reset them back to 755
webDriverAgentUrl	If provided, Appium will connect to an existing WebDriverAgent instance at this URL instead of starting a new one.	For example, http://localhost:8100

useNewWDA	If true, it forces the uninstall of any existing WebDriverAgent app on the device. This can provide stability in some situations; it defaults to false.	For example, `true`
wdaLaunchTimeout	Time, in ms, to wait for WebDriverAgent to be pingable; it defaults to `60000` ms.	For example, `30000`

Summary

In this chapter, we were introduced to the concept of desired capabilities and how they set the context of automation. We also got acquainted with the mandatory capabilities and the device-specific desired capabilities, such as Android and iOS. We also refactored the code to use some of these capabilities around launching the session using an app package name and app activity. We also refactored the code to launch the Android virtual device.

In the next chapter, we will build on a couple of more scenarios and explore Appium inspector. We will learn how to find locators and debug hybrid apps via the Chrome browser.

5
Understanding Appium Inspector to Find Locators

In the last chapter, we saw how to start Appium server programmatically and use the desired capabilities to set the context for test execution. We saw how to launch an emulator from code and invoke the test on an app that is preinstalled. We also took a look at detailed Android-specific capabilities and iOS-specific capabilities and refactored the test to use a few.

To write tests, we need to find the locators, and sometimes these locators are not easily available. We need to make our own locators using xPath in that case. In this chapter, we will take a look at the following:

- How to use Appium inspector to find locators
- How to use UI Automator Viewer
- How to use Chrome browser to debug mobile web apps

Appium inspector

We read about Appium inspector briefly in `Chapter 3`, *Writing Your First Appium Test*, to find out the element we want to click on. It's a handy tool for element discovery and understating the hierarchy. Let's take a thorough look at Appium inspector and the possibilities it opens for us. Let's launch the Appium inspector by following these steps:

1. Launch the emulator.
2. Ensure that the Quikr app is installed on the emulator.

3. Click on the **Android Settings** icon in the Appium app, select the **Package** as **com.quikr**, and select the **Launch Activity** as **com.quikr.old.SplashActivity**. (Refer Appendix to learn to find out Package Name and Launch Activity)

4. Select the **Platform Name** as **Android**, **Automation Name** as **Appium**, and the **Platform Version** as **5.1 Lollipop (API Level 22)**.

5. Check the **Device Name** box and enter **Nexus**.

6. Click on the **Launch** button.

7. Once the Appium console shows the following output, click on the **Appium Inspector** icon:

```
[HTTP] --> GET /wd/hub/status {}
[MJSONWP] Calling AppiumDriver.getStatus() with args: []
[MJSONWP] Responding to client with driver.getStatus() result:
{"build":{"version":"1.5.3"...
[HTTP] <-- GET /wd/hub/status 200 48 ms - 83
```

This will launch the **Appium Inspector** window, as follows:

We briefly discussed the Appium inspector window in `Chapter 2`, *Machine Setup*. Let's explore how to use this window to derive locators. Inspector loads this window with the UI element selected on the right pane. So, in most cases fields will have **resource-id**, which will contain the ID value of the element. It also shows the attributes of the element, which include **type**, **text**, **index**, **enabled**, and **location**. All these values can be used to derive the locator xPath to get a handle on the element. Let's navigate to the home page of the app by clicking on the **SKIP** link of the app on the emulator. Once done, click on the **Refresh** button in the Appium inspector window.

It will load this window:

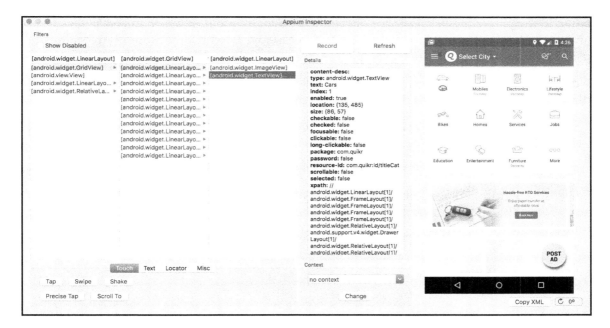

Now, assume the following test case:

```
Scenario: Search for a used Honda City car in Bangalore city

    When I launch Quikr app
    And I choose "Bangalore" as my city
    And I search for "Honda City" under Used Cars
    Then I should see the first car search result with "Honda"
```

For all the elements, we need locators. So, we need locators for the following for the preceding scenario:

1. Tap on **Select City.**
2. Type `Bangalore` in the **Search for your city** textbox.
3. Tap on the **Cars** category.
4. Type `Honda City` in the **Find a Car** textbox.
5. Tap on the **Find Used Cars** button.
6. Verify that the result is shown.

Now, in the preceding list, a couple of elements have unique IDs, which makes it easy to author the test, but some don't have IDs. In this case, we need to create our own locator using xpath. Let's run through a sample exercise in creating xpath for tapping on **Cars**.

Now, all the elements in the category list have the same ID, `titleCat`, but the text is different for each of them. Tap on the **Locator** tab in the Appium inspector window.

To form an xpath, we can use one of the following syntax:

```
xpath=//type[@attribute='value']
xpath=//type[contains(@attribute, 'value')]
```

So, let's try to form an xpath in this case. For **type**, we can use **android.widget.TextView** and for the attribute part, we need to use something that's unique. So, we have **text** as the unique value in this case:

Type: android.widget.TextView

Text: Cars

So, the xpath for **Cars** will be `//android.widget.TextView[@text='Cars']`, and this can be tested by clicking on the **Search** button in the window:

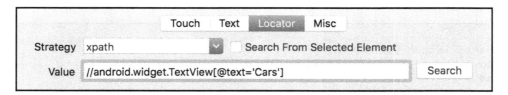

There's also an **xpath** attribute, which is present in the **Appium Inspector** window under the **Details** pane; the following is the value of xpath for the **Cars** element. The problem with this value is that it's not intuitive. The xpath value will work as long as Appium Inspector hierarchy and everything else remain the same:

```
xpath:
//android.widget.LinearLayout[1]/android.widget.FrameLayout[1]/android.widg
et.FrameLayout[1]/android.widget.FrameLayout[1]/android.widget.FrameLayout[
1]/android.widget.RelativeLayout[1]/android.support.v4.widget.DrawerLayout[
1]/android.widget.RelativeLayout[1]/android.widget.RelativeLayout[1]/androi
d.widget.FrameLayout[1]/android.widget.ScrollView[1]/android.widget.LinearL
ayout[1]/android.widget.GridView[1]/android.widget.LinearLayout[1]/android.
widget.TextView[1]
```

Implementing the other steps

Let's implement the preceding scenario. The first step to launch a quicker app is already automated, so let's figure out the dependency for the following step:

```
And I choose "Bangalore" as my city
```

1. Tap on **SKIP.**
2. Tap on **Select City.**
3. Enter `Bangalore` in the **search for your city** textbox.
4. Select the appearing value from the dropdown.

So, when you use the Inspector, you will notice that all the preceding elements have an ID, which can be easily used.

Here's the implementation for the same. We can create these methods in the **HomePageSteps** class file. To ensure that the click action has the element visible, we have used `Thread.sleep()`. We will refactor the same in a later chapter to use `WebDriver` wait:

```
@And("^I choose \"([^\"]*)\" as my city$")
public void iChooseAsMyCity(String city) throws Throwable {
    appiumDriver.findElement(By.id("skip")).click();

    Thread.sleep(2000);
    appiumDriver.findElement(By.id("citySpinner")).click();

    Thread.sleep(2000);
    appiumDriver.findElement(By.id("search_ET")).click();
    appiumDriver.findElement(By.id("search_ET")).sendKeys(city);
```

```
        Thread.sleep(2000);
        appiumDriver.findElement(By.id("city_name")).click();
}
```

Similarly, we can implement the other steps:

```
And I search for "Honda City" under Used Cars
Then I should see the first car search result with "Honda"
```

If you get the **Upgrade Available** popup, you can use the **Appium Inspector** window, as shown, and construct the **xpath** for the same:

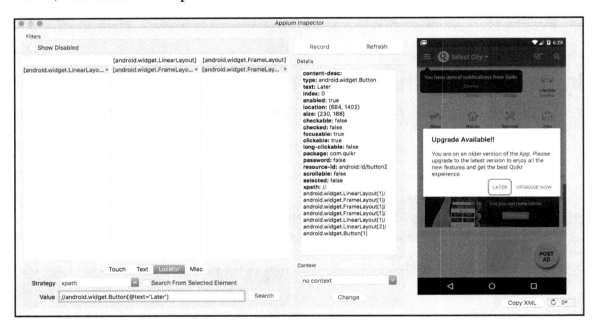

```
appiumDriver.findElement(By.xpath("//android.widget.Button[@text='Later']")
);
```

To implement the next step, which is `And I search for "Honda City" under Used Cars`, we need to perform the following steps:

1. Tap on the **Cars** category.

   Here's the code to do the same; the locator used is `xpath`:

   ```
   appiumDriver.findElement(By.xpath("//android.widget.
   TextView[@text='Cars']")).click();
   ```

2. Tap on the **Find a Car** textbox.

   This is the code to do the same; the locator used is `id`:

   ```
   appiumDriver.findElement(By.id("cnb_hp_choose_et")).click();
   ```

3. Type in `Honda City` in the textbox.

   The following is the code to do the same; the locator used is `id`:

   ```
   appiumDriver.findElement(By.id("cnb_search_text_et"))
   .sendKeys("Honda City");
   ```

4. Select the matching item from the results.

The following is the code to do the same; the locator used is **id**. To find the result, we need to click on one result from the list of results. So, we will be using the `findElements()` API, which will return a list of results and here we will query for the result we want:

```
appiumDriver.findElements(By.id("text1"));
```

To filter the result we want, we will iterate through the list and check for the element that contains the text **Honda City**. Here's the code for the same:

```
List<WebElement> results = appiumDriver.findElements(By.id("text1"));
for (WebElement result : results) {
    if (result.getText().contains(carName)){
        result.click();
        break;
    }
}
```

5. Tap on **Find Used Cars**.

   Here's the code to do the same; the locator used is **id**:

   ```
   appiumDriver.findElement(By.id("cnb_search_button")).click();
   ```

   So, the complete method will look like this:

```
@And("^I search for \"([^\"]*)\" under Used Cars$")
 public void iSearchForUnderUsedCars(String carName) throws Throwable {
appiumDriver.findElement(By.xpath("//android.widget.TextView[@text='Cars']"
)).click();
    appiumDriver.findElement(By.id("cnb_hp_choose_et")).click();
appiumDriver.findElement(By.id("cnb_search_text_et")).sendKeys(carName);
```

```
List<WebElement> results = appiumDriver.findElements(By.id("text1"));
for (WebElement result : results) {
    if (result.getText().contains(carName)) {
        result.click();
        break;
    }
}
appiumDriver.findElement(By.id("cnb_search_button")).click();
}
```

The last step to be automated is as follows:

```
Then I should see the first car search result with "Honda"
```

This is the code to do the same, and the locator used is **id**:

```
appiumDriver.findElements(By.id("cars_ad_list_title_tv"));
```

In the next code, we are checking that the header of each result item contains **Honda** as we have searched for **Honda City** car:

```
@Then("^I should see the first car search result with \"([^\"]*)\"$")
public void iShouldSeeTheFirstCarSearchResultWith(String arg0) throws
Throwable {
    List<WebElement> elements =
appiumDriver.findElements(By.id("cars_ad_list_title_tv"));
    Assert.assertTrue("Verified first result contains
Honda",elements.get(0).getText().startsWith(arg0));
}
```

So, this completes the implementation of the new scenario. You would have noticed that, to use Appium inspector, we need to start the session from scratch. There are times when we perform certain transactions and steps to arrive on a screen and then see the locators. There is a direct way to check for locators without having to use Appium.

UI Automator Viewer

There is an alternate way of just seeing locators using `uiautomatorviewer`, bundled by Android SDK. If you have set up the Android SDK path, open the Terminal (Command Prompt in Windows) and type in the `uiautomatorviewer` command. This will launch a blank window with a couple of icons on top, as illustrated in the following screenshot. It is present under the **tools** folder in Android SDK:

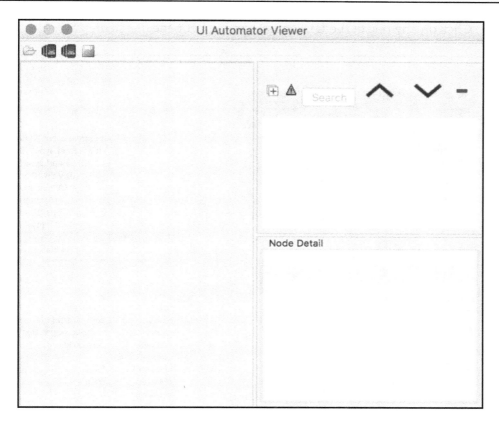

We use this tool to find out the application UI hierarchy and show the details of the elements present in the UI. We can inspect the attributes of an element by clicking on the element.

Steps to use UI Automator Viewer:

1. Prerequisites: Emulator is running and the Android SDK path is set.
2. Open Terminal and type in `uiautomatorviewer` (for Windows, open Command Prompt and type in `uiautomatorviewer`).
3. Launch the app under test in the emulator, Quikr in our case.
4. Click on the second icon, which is **Device Screenshot**.
5. The **UI Automator Viewer** window will launch, with the following screen.

6. Click on any one of the UI elements on the left pane:

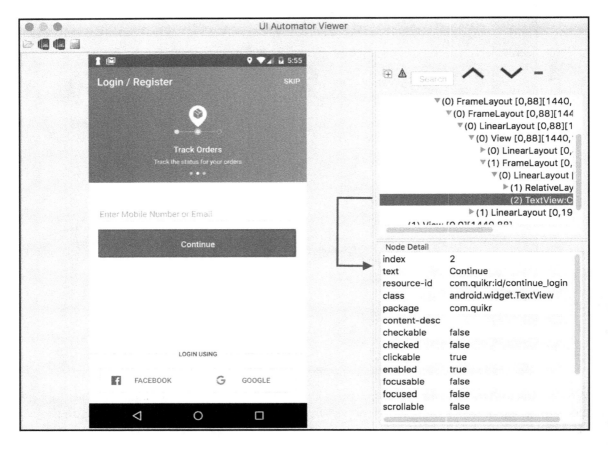

7. In the right pane, we see two things: UI hierarchy and the element details, such as **index**, **resource-id**, and **class**.
8. Clicking on the **Save** icon allows us to save the screenshot and the XML layout of the screen with all the node details.

This is a lightweight way to check for locators without going through the process of starting an Appium session. It can be invoked on any screen of the emulator.

Debugging mobile web apps using Chrome Inspect

The Chrome browser comes with a lot of handy features under Dev Tools. It can be used to debug and profile a mobile web app. Chrome Dev Tools can be launched from **More Tools > Developer Tools** under the Chrome menu:

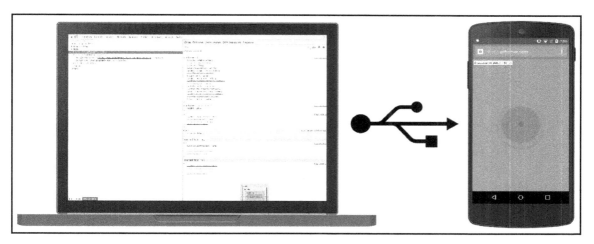

To use the Chrome inspect feature, we need to be running the emulator/device upward of **Android 4.0** and **Chrome for Android** has to be installed on the emulator/device.

To use the device, we need to enable certain settings: on the Android device, select **Settings > Developer Options > Enable USB Debugging. Developer Options** is hidden, by default, on Android 4.2 and later. We will take a detailed look at it when we move on to devices for test execution. For now, emulators will show up by default and do not require any permission settings.

Carry out the following steps to use the Chrome Inspect feature:

1. Launch the emulator and open the mobile web app on the Chrome browser of the emulator.
2. Open Chrome browser on your mobile and type in **chrome://inspect**.
3. It displays a list of debug-enabled web views on the emulator running (or the device connected).

4. This will load the following details, showing the remote device connected, which in this case is an emulator running on host **192.168.56.101:5555**:

5. The emulator is running Chrome browser version **55.0.2883.91**. It also shows the site opened along with the URL and certain options available, such as **inspect**, **focus tab**, **reload**, and **close**.

6. To start debugging, click on the **inspect** link, and it will open the **Developer Tools** window.

7. The screen is divided into three sections: one displaying the **mobile UI**, the second one with **element/DOM hierarchy** highlighted, and the last one showing **Styles** and **Event Listeners**.

8. Overall, there are nine main tools available: **Elements, Console, Sources, Network, Timelines, Profiles, Application, Security**, and **Audits**, as shown:

In the preceding screenshot, the transparent portions represent device interfaces, such as the **Chrome omnibox** or the **Android status** bar.

Once we have this window open, we can click on the **Select Element** icon and then click on the element in the left pane whose locator we want to find. We can also click on **Toggle Screencast** to view the content of the device/emulator in the DevTools instance. When in toggled mode, the icon color will change to blue . The screen on the left can be interacted with using clicks, which are generally translated into tap actions, and key strokes are sent to the device.

Summary

In this chapter, we delved into how to use Appium Inspector to find locators of UI elements. Also, we learned to derive xpath over the Appium-generated xpath, which is long and difficult to comprehend and maintain. Xpath has to be used if there are no readily available locators, preferably IDs. We also implemented a few cucumber steps to use locators and learned to select the element we need from the list.

We learned about the UI Automator Viewer and how to use it. We also learnt about debugging mobile web apps or webviews using Chrome browser's inspect feature.

6

How to Synchronize Tests

In the earlier chapters, we completed the journey of writing a basic test scenario that runs on the emulator. We started with the machine setup, creating an Appium Java project and then writing the first Appium test. We also looked at how to use the Appium inspector to find locators. During this process, we wrote a couple of scenarios and automated them. Robustness and reliability are the traits of a good automated test. However, while writing a test, sometimes we need to keep the test execution speed in sync with the actual app performance; this way, the script won't fail for issues such as the app not loading rapidly. So far, we handled it using `Thread.sleep()` in our code, which is not the best way to handle synchronization.

In this chapter, we will learn about the following:

- Different driver types available in Appium
- Wait strategies:
    - Implicit wait
    - Explicit wait
    - Fluent wait

And we will refactor the code to implement these.

AppiumDriver

If you refer to `Chapter 3`, *Writing Your First Appium Test,* and remember the boilerplate code generated, it creates an instance `AppiumDriver`:

```
wd = new AppiumDriver(new URL("http://0.0.0.0:4723/wd/hub"), capabilities);
```

Let's take some time to understand what types of driver Appium allows us to create.

Certainly, `AppiumDriver` was generated by the boilerplate code. Let's take a look at the other drivers:

- `RemoteWebDriver`: It comes from Selenium. It has two components: a server and a client. A server is a component that listens on a port for various requests from the client. The client translates the script to the JSON payload and sends it to the server using the JSON wire protocol.
- `AppiumDriver`: It inherits from the `RemoteWebDriver` and adds functions that are handy for mobile automation. It can be used to automate both Android and iOS apps; however, it lacks device family-specific functions. The direct subclasses are `AndroidDriver`, `IOSDriver`, and `WindowsDriver`.
- `AndroidDriver`: It inherits from `AppiumDriver` and adds in additional functions that are highly contextual to the family of Android devices for automation. If you are working only on an Android project, then it's highly recommended for you to use this driver.
- `IOSDriver`: It inherits from and adds in additional functions that are highly contextual to the family of iOS devices for mobile automation. If you are working on iOS app automation, then it's highly recommended for you to use this driver:

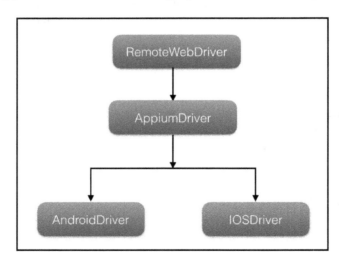

Understanding the different types of driver is important as there are different methods available specifically for certain driver types, and they can help you to solve the problem of writing a lot of explicit code. All you need to do is typecast the driver and use that method. We will explore some gestures supported only for specific drivers in the next chapter.

Implicit wait

Implicit wait is a way to tell the Appium driver to poll the DOM (Document Object Model) for a certain amount of time before throwing an exception to the effect that it can't find the element on the page. The default timeout value is set to 0 seconds. Once we set the implicit wait to a specified time, it persists for the life of the `webdriver` object instance. How to set an implicit wait is explained here:

```
appiumDriver.manage().timeouts().implicitlyWait(10, TimeUnit.SECONDS);
```

So, what this implies is letting the driver instance wait for a maximum of 10 seconds before throwing the `NoSuchElement` exception. We need to be watchful about the implicit usage. The Appium boilerplate generally gives us the code with the implicit wait implementation, so note the preceding line in the `HomePageWebSteps` class file, as shown:

```
appiumDriver = new AppiumDriver(new URL("http://0.0.0.0:4723/wd/hub"),
capabilities);
appiumDriver.manage().timeouts().implicitlyWait(30, TimeUnit.SECONDS);
```

Increasing the implicit wait timeout should be used judiciously as it will have an adverse effect on the overall test execution time, especially when used with slower locator strategies, such as `xpath`.

This just removes a lot of indeterministic wait from the code. Implicit wait is most suited when there is a variation in app response time due to network speed.

Explicit wait

There are times when the app under test can be slow on certain specific elements, such as page submit, form submit, or somewhere it fetches data from an external system and takes a little more time to load. In that case, using implicit wait to handle the situation will be a flawed approach, given that it has to wait for each and every element for the same specified time.

To handle this situation, we can use explicit wait for such elements. In explicit wait, we tell the web driver instance to wait for a certain condition invoked through `ExpectedConditions`. So, this wait applies explicitly to the specified element. Explicit wait can be invoked using this code:

```
WebDriverWait wait = new WebDriverWait(appiumDriver, 10);
wait.until(ExpectedConditions.visibilityOfElementLocated(By.id("text1")));
```

In the preceding code, we are creating an instance of `WebDriverWait` with a maximum waiting time of 10 seconds and then using an `ExpectedConditions`, which tells the driver to wait till the visibility of the specified element can be located. `ExpectedConditions` has a bunch of methods available to be used under different conditions. `WebDriverWait`, by default, calls `ExpectedConditions` every 500 ms until it returns successfully, otherwise it throws the `TimeoutException`, as follows:

```
org.openqa.selenium.TimeoutException: Expected condition failed: waiting
for visibility of element located by By.id: text1 (tried for 10 second(s)
with 500 MILLISECONDS interval)
```

While automating, you will typically need the given conditions to be met for an element and, for each of the following, `ExpectedConditions` provides a set of predefined conditions:

- Web element is present and clickable
- Web element is selected
- Web element is invisible
- Selected web element
- Presence of web element located by
- Wait for a particular condition
- Text present in a web element

Here's a list of all the methods available under `ExpectedConditions`:

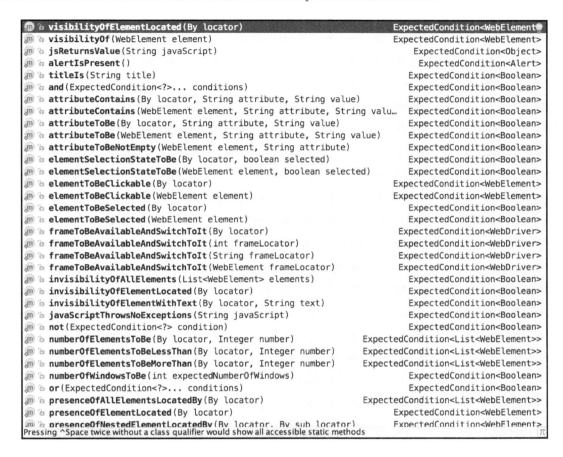

Explicit wait is also used to check a specified property of an element, such as visibility, click-ability, invisibility, and selection state.

Let's go ahead and refactor some of the code to introduce explicit wait. Some of the places to add explicit wait, would be where we are typing in a **textbox** to search for an item. Consider the following examples:

- Choosing my city:

```
WebDriverWait wait = new WebDriverWait(appiumDriver, 10);
wait.until(ExpectedConditions.visibilityOfElementLocated
(By.id("city_name")));
appiumDriver.findElement(By.id("city_name")).click();
```

- Searching for a specified car:

```
WebDriverWait wait = new WebDriverWait(appiumDriver, 10);
wait.until(ExpectedConditions.visibilityOfElementLocated
(By.id("text1")));
List<WebElement> results =
appiumDriver.findElements(By.id("text1"));
```

So there are two different methods we are adding explicit wait to; we can even set a different timeout for each element. However, one observation would be that the code is repeating. We will come back to refactor this piece in subsequent chapters to create something called a base page class, which hosts all such commonly used methods that can be used in each page class.

Make the preceding changes in your code and run the test; the test should run smoothly, as earlier. Here's an implementation in one of the methods:

```
@And("^I choose \"([^\"]*)\" as my city$")
public void iChooseAsMyCity(String city) throws Throwable {
    WebDriverWait wait = new WebDriverWait(appiumDriver, 10);
    wait.until(ExpectedConditions.visibilityOfElementLocated
    (By.id("skip")));
    appiumDriver.findElement(By.id("skip")).click();

    try {
        if (appiumDriver.findElement(By.xpath
        ("//android.widget.Button[@text='Later']")).isDisplayed())
            appiumDriver.findElement(By.xpath
            ("//android.widget.Button[@text='Later']")).click();
    } catch (Exception e) {
        //Do nothing
    }

    appiumDriver.findElement(By.id("citySpinner")).click();
    appiumDriver.findElement(By.id("search_ET")).click();
    appiumDriver.findElement(By.id("search_ET")).sendKeys(city);

    wait.until(ExpectedConditions.visibilityOfElementLocated
    (By.id("city_name")));
    appiumDriver.findElement(By.id("city_name")).click();
}
```

Let's move on to understand a wait type that is somewhat more specific and lets us further customize it.

Fluent wait

Fluent wait is a type of explicit wait where we can define polling intervals and ignore certain exceptions to proceed with further script execution even if the element is not found.

So, when we specify a fluent wait, we provide the following:

- Maximum wait time
- Polling interval or frequency to check the element
- Any specific exception to ignore
- Message that should appear after timeout

A simple example of a fluent wait implementation is as follows:

```
Wait wait = new FluentWait(appiumDriver)
        .withTimeout(10, TimeUnit.SECONDS)
        .pollingEvery(250, TimeUnit.MILLISECONDS)
        .ignoring(NoSuchElementException.class)
        .ignoring(TimeoutException.class);

wait.until(ExpectedConditions.visibilityOfElementLocated
(By.id("text1")));
```

Let's implement the preceding in the `iChooseAsMyCity(String city)` method and re-run the test to see what the results are:

```
@And("^I choose \"([^\"]*)\" as my city$")
public void iChooseAsMyCity(String city) throws Throwable {
    Wait wait = new FluentWait(appiumDriver)
            .withTimeout(10, TimeUnit.SECONDS)
            .pollingEvery(250, TimeUnit.MILLISECONDS)
            .ignoring(NoSuchElementException.class)
            .ignoring(TimeoutException.class);
    wait.until(ExpectedConditions.visibilityOfElementLocated
    (By.id("skip")));

    appiumDriver.findElement(By.id("skip")).click();

    try {
        if (appiumDriver.findElement(By.xpath
        ("//android.widget.Button[@text='Later']")).isDisplayed())
            appiumDriver.findElement(By.xpath
            ("//android.widget.Button[@text='Later']")).click();
    } catch (Exception e) {
        //Do nothing
    }
```

```
    appiumDriver.findElement(By.id("citySpinner")).click();
    appiumDriver.findElement(By.id("search_ET")).click();
    appiumDriver.findElement(By.id("search_ET")).sendKeys(city);

    wait.until(ExpectedConditions.visibilityOfElementLocated
    (By.id("city_name")));
    appiumDriver.findElement(By.id("city_name")).click();
}
```

Run this test, the result is the same as the earlier ones. Basically, all the three approaches handle the element wait in different ways and give us the same result. However, we need to choose the most suitable approach based on the situation.

Summary

In this chapter, we learned to implement wait strategy using implicit a wait, explicit wait, and fluent wait. We also learned how these waits are different and in which way. We also learned the predefined conditions that `ExpectedConditions` allows one to use. We modified some of the tests to run them using the new wait strategies and saw that all of them work seamlessly.

In the next chapter, we will see how to automate gestures, such as tap, long press, swipe, and scroll. We will also refactor the existing test to organize it in a much easier to maintain structure.

7
How to Automate Gestures

In the earlier chapters, we learned how to set up and write a basic Appium test. We started with a scenario and learned how to use the Appium inspector and write a few automated tests. We also learned the concept of desired capabilities and saw how to use them. We learned how to add synchronization in tests and the different types of wait strategy Appium allows us to use. In this chapter, we will learn how to automate different gestures, such as the following:

- Tap
- Swipe
- Drag
- Scroll to
- Slider
- Shake
- Long tap
- Orientation

Let's start with each of the afore mentioned and learn its implementation and details.

Gestures

Mobile devices allow a multitude of gestures, which can be used across the app. However, there are no standards as to what gestures an app must implement. Some of the gestures most typically used are tap, swipe, pinch, and double tap. One good thing with mobiles is that these gestures are constantly evolving and eventually become natural to use. So, let's take a look at the different gestures and how they can be implemented.

TouchAction

Appium implements the new **TouchAction** API, which allows chaining touch events and, thereby, facilitates gesture implementation. Touch Action is pretty robust and supports a multitude of gestures, which ease the simulation:

```
m  🔒 perform()                                              TouchAction
m  🔒 moveTo(WebElement el, int x, int y)                    TouchAction
m  🔒 press(int x, int y)                                    TouchAction
m  🔒 release()                                               TouchAction
m  🔒 press(WebElement el)                                   TouchAction
m  🔒 cancel()                                                      void
m  🔒 longPress(int x, int y)                                TouchAction
m  🔒 longPress(int x, int y, int duration)                  TouchAction
m  🔒 longPress(WebElement el)                               TouchAction
m  🔒 longPress(WebElement el, int duration)                 TouchAction
m  🔒 longPress(WebElement el, int x, int y)                 TouchAction
m  🔒 longPress(WebElement el, int x, int y, int duration)   TouchAction
m  🔒 moveTo(int x, int y)                                   TouchAction
m  🔒 moveTo(WebElement el)                                  TouchAction
m  🔒 press(WebElement el, int x, int y)                     TouchAction
m  🔒 tap(int x, int y)                                      TouchAction
m  🔒 tap(WebElement el)                                     TouchAction
m  🔒 tap(WebElement el, int x, int y)                       TouchAction
m  🔒 waitAction()                                           TouchAction
m  🔒 wait()                                                        void
m  🔒 wait(long timeout)                                            void
m  🔒 wait(long timeout, int nanos)                                 void
m  🔒 waitAction(int ms)                                     TouchAction
m  🔒 equals(Object obj)                                          boolean
```

We will discuss some of the methods mentioned earlier, that **TouchAction** supports:

- **press**:
  - `press(WebElement el)`: This method allows you to press on the center of the element
  - `press(int x, int y)`: This method allows you to press on an absolute position (x and y coordinates)

- `press(WebElement el, int x, int y)`: This method allows you to press on an element offset from the upper-left corner by a number of pixels:

Consider the following usage examples:

```
TouchAction action = new TouchAction(appiumDriver);
action.press(appiumDriver.findElement(By.id("valid_id"))).perform();
Point point =appiumDriver.findElementById("valid_id").getLocation();
new TouchAction(appiumDriver).press(point.x + 20, point.y +
30).waitAction(1000).release().perform();
```

- **release**:
    - `release()`: This method withdraws the touch

Consider the following example:

```
action.release();
```

- **long press**:
    - `longPress(int x, int y)`: This method allows you to press and hold the absolute position x,y until the context event has fired
    - `longPress(int x, int y, int duration)`: This method allows you to press and hold for the specified duration at an absolute position x,y until the context event has fired
    - `longPress(WebElement el)`: This method allows you to press and hold the center of an element until the context event has fired

- `longPress(WebElement el, int duration)`: This method allows you to press and hold the center of an element until the context event has fired
- `longPress(WebElement el, int x, int y)`: This method allows you to press and hold the elements in the upper-left corner, offset by the x,y amount, until the context event has fired:

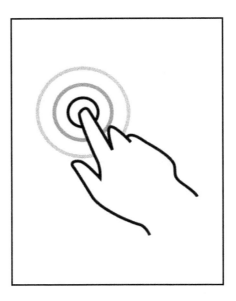

Consider the following usage example:

```
TouchAction action = new TouchAction(appiumDriver);
action.longPress(appiumDriver.findElement(By.id("valid_id"))).p
erform();
```

- **move**:
    - `moveTo(int x, int y)`: This method allows you to move the current touch to a new position that is relative to the current position
    - `moveTo(WebElement el)`: This method allows you to move the current touch to the center of the specified element
    - `moveTo(WebElement el, int x, int y)`: This method allows you to move the current touch to the specified element and offset from the upper-left corner:

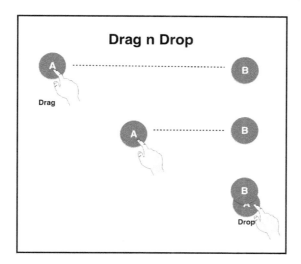

Consider this usage example:

```
WebElement drag = appiumDriver.findElement(By.id("drag_1"));
WebElement drop = appiumDriver.findElement(By.id("drop_1"));
TouchAction dragNDrop = new
TouchAction(appiumDriver).longPress(drag).moveTo(drop).release(
);
dragNDrop.perform();
```

- **perform**:
  - `perform()`: This allows you to perform a chain of actions
- **tap**:
  - `tap(int x, int y)`: This method allows you to tap an absolute position (x,y) on the screen
  - `tap(WebElement el)`: This method allows you to tap the center of the specified element
  - `tap(WebElement el, int x, int y)`: This method allows you to tap an element with the specified offset from the upper-left corner
  - `waitAction()`: This method allows you to wait for the action to be completed and is used as a no-operation in multi-chaining
  - `waitAction(int ms)`: This method allows you to wait for a specified amount of time to pass before it continues with performing the next touch action

Consider this usage example:

```
TouchAction action = new TouchAction(appiumDriver);
action.tap(appiumDriver.findElement(By.id("valid_id"))).perform();
// A case where an UI element is Start and when pressed enables
another element called Stop
Point center1 =
appiumDriver.findElementById("id_start").getCenter();

TouchAction action = new TouchAction(appiumDriver)
        .tap(center1.x, center1.y)
        .tap(appiumDriver.findElementById("id_stop"),5,5);
action.perform();
```

In the preceding API, when we pass coordinates along with the web element, the coordinates are treated as relative to the web element position. When we call the `perform` method, the sequence of the constructed event is sent to Appium, and the touch action is performed on the device.

MultiTouch

Appium gives you the option to construct a MultiTouch action by chaining touch actions. So, we can chain all the actions that the `TouchAction` class supports. MultiTouch is a collection of TouchActions and allows two operations: add and perform:

- `Add` is used to chain another TouchAction
- `Perform` is called to send all TouchActions to Appium in the same order

Let's take a look at its usage:

```
TouchAction action1 = new TouchAction(appiumDriver).tap(webElement1);

TouchAction action2 = new TouchAction(appiumDriver).tap(webElement2);

MultiTouchAction multiTouchAction = new MultiTouchAction(appiumDriver);
multiTouchAction.add(action1).add(action2).perform();
```

Scroll

One of the most commonly used gestures on mobiles is scroll. Earlier, there were two methods available for scrolling: `scrollTo(String text)` and `ScrollToExact(String text)`. However, recently, both functions have been deprecated. To solve this, we can use swipe to perform the scroll functions and pass in parameters that are based on the relative height and width of the screen.

To scroll down, use the following code snippet, where we have fixed the x component and moved the y component:

```
public void scrollDown() {
    int height = driver.manage().window().getSize().getHeight();
    androidDriver.swipe(5, height * 2 / 3, 5, height / 3, 1000);
}
```

To scroll up, use this code snippet where we have fixed the x component and moved the y component:

```
public void scrollUp() {
    int height = driver.manage().window().getSize().getHeight();
    androidDriver.swipe(5, height / 3, 5, height * 2 / 3, 1000);
}
```

Once we have the preceding implementation, we can implement scroll down to an element using the following code. Change the `System.out` to the appropriate assertion.

```
public void scrollDownTo(By locatorOfElement) {
    int i = 0;
    while (i < 12) {
        if (driver.findElements(locatorOfElement).size() > 0)
            return;
        scrollDown();
        i++;
    }
    System.out.println("Couldn't find element: " +
locatorOfElement.toString());
}
```

We can even implement a scroll down to text on similar implementations; refer to this code snippet:

```
public void scrollDownTo(String text) {
    By locatorOfElement = By.xpath("//*[@text=\"" + text + "\"]");
    androidDriver.hideKeyboard();
    int i = 0;
    while (i < 12) {
        if (androidDriver.findElements(locatorOfElement).size() > 0)
            return;
        scrollDown();
        i++;
    }
    System.out.println("Couldn't find text : " +
locatorOfElement.toString());
}
```

The preceding code can be used for reference and for developing your own customized scroll method.

Swipe

Swipe is another commonly used gesture on a mobile device. The swipe functionality is relative to the device height and width. So, we can use the relative device height and width to implement the swipe functionality. Android Driver supports the swipe method, and we can use it:

```
void swipe(int startx, int starty, int endx, int endy, int duration);
```

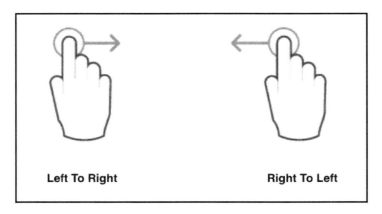

Here's the code snippet for swiping from left to right:

```
public void swipeLeftToRight() {
    int height = driver.manage().window().getSize().getHeight();
    int width = driver.manage().window().getSize().getWidth();
    androidDriver.swipe(width/3, height/2, (width*2)/3, height/2, 100);
}
```

Similarly, we can implement right to left:

```
public void swipeRightToLeft() {
    int height = driver.manage().window().getSize().getHeight();
    int width = driver.manage().window().getSize().getWidth();
    androidDriver.swipe((width*9)/10, height/2, width/10, height/2, 1000);
}
```

In a similar format, we can even implement swiping from one element to another element, based on the element location:

```
public void swipeFromTo(WebElement startElement, WebElement stopElement) {
    androidDriver.swipe(startElement.getLocation().getX(),
    startElement.getLocation().getY(),
    stopElement.getLocation().getX(), stopElement.getLocation().getY(),
    1000);
}
```

One thing to note here is that the swipe method is made available as part of `androidDriver` and not `appiumDriver`; hence, we need to type cast the driver to `androidDriver` in the current implementation and then use this method.

Orientation

Many times, we need to change the device orientation to see a different view or to perform some other action/assertion. Appium exposes a method to change the orientation from landscape to portrait and vice versa:

```
rotate(ScreenOrientation orientation)
```

`ScreenOrientation` gives a possible screen orientation and supports `LANDSCAPE` or `PORTRAIT`.

Using the rotate API:

```
androidDriver.rotate(ScreenOrientation.LANDSCAPE);
```

Appium also exposes an API to get the current orientation of the device:

```
androidDriver.getOrientation()
```

A good practice is to perform the operation and change the orientation back so that it doesn't affect the other test cases unless it is intended to do so.

Summary

In this chapter, we learned how to automate frequently-used gestures, such as press, long press, touch, and MultiTouch. We learned the APIs exposed by Appium and also the usage of those APIs. These code snippets can be used to implement different touch functionalities in your app. We also learned about device orientation and how to toggle it between the portrait and landscape modes. We also discussed the fact that some of these methods are available only to androidDriver.

In the next chapter, we will learn about design patterns in test automation and delve into one of the most popular design patterns. We will also learn about some of the best practices for framework designing.

8

Design Patterns in Test Automation

In the past chapters, we learned about gestures and how to implement gestures in mobile automation. Until now, we have learned almost all the major aspects of Appium, right from understanding the app to writing a basic test in cucumber and automating it. If you notice the code we have written, we can see elements of repeatability and lack of structure and design in the whole approach. There are a couple of design patterns that are used in test automation framework, and some of them are s, Singletons, Facades, Strategy design patterns, and so on.

In this chapter, we will take a look at the very popular and widely used design pattern as well as the most fundamental aspect of test automation which is assertion:

- Page Object pattern
- Implementing assertions

Before we get on to the concept of the Page Object pattern, let's do one more round of refactoring and introduce the concept of `BaseSteps`, the same is illustrated below.

We will implement this in the current state of automation to give it a more structured and organized look. Let's start with understanding the Page Object pattern concept.

Refactor -1

Let's recall the `HomePageSteps` class and the `iLaunchQuikrApp()` method:

```
@When("^I launch Quikr app$")
 public void iLaunchQuikrApp() throws Throwable {
     DesiredCapabilities capabilities = new DesiredCapabilities();
     capabilities.setCapability("platformName", "Android");
     capabilities.setCapability("platformVersion", "5.1");
     capabilities.setCapability("deviceName", "Nexus");
     capabilities.setCapability("noReset", false);
     capabilities.setCapability("fullReset", true);
     capabilities.setCapability("app",
"/Users/nishant/Development/HelloAppium/app/quikr.apk");

     appiumDriver = new AppiumDriver(new URL("http://0.0.0.0:4723/wd/hub"),
capabilities);
     appiumDriver.manage().timeouts().implicitlyWait(60, TimeUnit.SECONDS);
 }
```

Now, the instance of `appiumDriver` can be used by other step classes and not only this one. So, to solve this, we will declare a `BaseSteps` class, which creates the `AppiumDriver` instance to be used throughout the test session.

Follow the steps below:

1. Select the following line in the `iLaunchQuikrApp()` method and click on **Refactor > Extract > Superclass...** :

   ```
   private AppiumDriver appiumDriver;
   ```

2. Enter `BaseSteps` as the **Super class name**, select **appiumDriver:AppiumDriver** under **Member**, and click on **Refactor**:

3. This will be the generated code for the same:

```
package steps;

import io.appium.java_client.AppiumDriver;

public class BaseSteps {
  protected AppiumDriver appiumDriver;
}
```

Note that the `AppiumDriver` access modifier changed from `private` to `protected`. This makes it accessible within the package and outside the package, but through inheritance only.

4. Add a `static` keyword to the `appiumDriver` instance as we want the same instance to persist for the session run:

```
protected static AppiumDriver appiumDriver;
```

5. Let's open the `StartingSteps` class and modify it a bit. We also need to move the `DesiredCapabilities` section from the preceding `iLaunchQuikrApp()` method to the `StartingSteps` class:

```
@Before
public void startAppiumServer() throws IOException {

int port = 4723;
    String nodeJS_Path = "C://Program Files//NodeJS//node.exe";
    String appiumJS_Path = "C://Program
Files//Appium//node_modules//appium//bin//appium.js";

    String osName = System.getProperty("os.name");

if (osName.contains("Mac")) {
appiumService = AppiumDriverLocalService.buildService(new
AppiumServiceBuilder()
                .usingDriverExecutable(new
File(("/usr/local/bin/node")))
                .withAppiumJS(new File(("/usr/local/bin/appium")))
                .withIPAddress("0.0.0.0")
                .usingPort(port)
                .withArgument(GeneralServerFlag.SESSION_OVERRIDE)
                .withLogFile(new File("build/appium.log")));
    } else if (osName.contains("Windows")) {
appiumService = AppiumDriverLocalService.buildService(new
AppiumServiceBuilder()
                .usingDriverExecutable(new File(nodeJS_Path))
                .withAppiumJS(new File(appiumJS_Path))
                .withIPAddress("0.0.0.0")
                .usingPort(port)
                .withArgument(GeneralServerFlag.SESSION_OVERRIDE)
                .withLogFile(new File("build/appium.log")));
    }
appiumService.start();

    DesiredCapabilities capabilities = new DesiredCapabilities();
    capabilities.setCapability("platformName", "Android");
    capabilities.setCapability("platformVersion", "5.1");
    capabilities.setCapability("deviceName", "Nexus6");
    capabilities.setCapability("noReset", false);
    capabilities.setCapability("fullReset", true);
```

```
        capabilities.setCapability("app",
    "/Users/nishant/Development/HelloAppium/app/quikr.apk");

    appiumDriver = new AppiumDriver(new
    URL("http://0.0.0.0:4723/wd/hub"), capabilities);
    appiumDriver.manage().timeouts().implicitlyWait(10,
    TimeUnit.SECONDS);
    }
```

6. Also, in the teardown method, we need to add `driver.quit()`, which will close the session before stopping the Appium server:

```
@After
public void closeAppiumServerSession() {
        appiumDriver.quit();
        appiumService.stop();
}
```

7. So, the `StartingSteps` class should look as shown:

```
public class StartingSteps extends BaseSteps {

private AppiumDriverLocalService appiumService;


@Before
public void startAppiumServer() throws IOException {

int port = 4723;
    String nodeJS_Path = "C://Program Files//NodeJS//node.exe";
    String appiumJS_Path = "C://Program
Files//Appium//node_modules//appium//bin//appium.js";

    String osName = System.getProperty("os.name");

if (osName.contains("Mac")) {
appiumService = AppiumDriverLocalService.buildService(new
AppiumServiceBuilder()
                .usingDriverExecutable(new
File(("/usr/local/bin/node")))
                .withAppiumJS(new File(("/usr/local/bin/appium")))
                .withIPAddress("0.0.0.0")
                .usingPort(port)
                .withArgument(GeneralServerFlag.SESSION_OVERRIDE)
                .withLogFile(new File("build/appium.log")));
    } else if (osName.contains("Windows")) {
appiumService = AppiumDriverLocalService.buildService(new
AppiumServiceBuilder()
```

```
                            .usingDriverExecutable(new File(nodeJS_Path))
                            .withAppiumJS(new File(appiumJS_Path))
                            .withIPAddress("0.0.0.0")
                            .usingPort(port)
                            .withArgument(GeneralServerFlag.SESSION_OVERRIDE)
                            .withLogFile(new File("build/appium.log"))));
        }
    appiumService.start();

        DesiredCapabilities capabilities = new DesiredCapabilities();
        capabilities.setCapability("platformName", "Android");
        capabilities.setCapability("platformVersion", "5.1");
        capabilities.setCapability("deviceName", "Nexus6");
        capabilities.setCapability("noReset", false);
        capabilities.setCapability("fullReset", true);
        capabilities.setCapability("app",
    "/Users/nishant/Development/HelloAppium/app/quikr.apk");

    appiumDriver = new AppiumDriver(new
    URL("http://0.0.0.0:4723/wd/hub"), capabilities);
    appiumDriver.manage().timeouts().implicitlyWait(10,
    TimeUnit.SECONDS);
    }

    @After
    public void closeAppiumServerSession() {
    appiumDriver.quit();
    appiumService.stop();
        }
    }
```

8. Now, the `iLaunchQuikrApp()` method becomes empty after moving all the code to the `StartingSteps` class. We can use this method to perform some assertion, or we can rename it to serve another purpose. For now, we will just add some checkpoints to it:

```
@When("^I launch Quikr app$")
public void iLaunchQuikrApp() throws Throwable {
appiumDriver.findElement(By.id("login_register_view")).isDisplayed(
);
}
```

Now, we have moved all the infrastructure code to `StartingSteps` and our `HomePageSteps` contains only methods that perform actions on the home page of the app.

Let's take a look at the design pattern to further organize and structure the code.

Page Object pattern

Here are some important aspects of a good framework design, which we tend to base our decision on:

- Avoiding duplication of code
- Tests should be more readable
- Tests should be easy to maintain
- Accommodating changes should be easy
- Enhanced reliability
- A structure that is easy to scale with the growth of the project

Page Object pattern is about modelling your app's UI as an object. A Page Object wraps the UI of a page with an app-specific API, which allows us to manipulate page elements. Let's understand the same with respect to the following image. The following page serves the purpose of both login and registration. It's the first page that gets displayed when we launch the app; for the sake of our conversation, let's call this a landing page. The page contains UI elements such as skip, mobile number text field, continue button, and Facebook and Google sign in buttons:

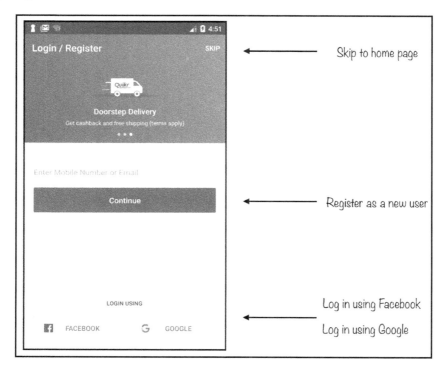

When we apply the Page Object concept, the preceding page will typically perform the following services for any user using the app:

- **Skip to home page**
- **Register as a new user**
- **Log in using Facebook**
- **Log in using Google**

In this case, the Page Object has the complete knowledge of the page elements and the services it can perform on that page. When we model a page this way, it can hide the UI elements from the consumer of the page and expose only the service one can perform on that UI element via accessor methods. Let's create a sample page class for the shown page:

```
public class LandingPage {
    AppiumDriver appiumDriver;

    @FindBy(id = "skip")
    private WebElement skipLink;

    @FindBy(id = "login_register_view")
    private WebElement mobileOrEmailField;

    @FindBy(id = "continue_login")
    private WebElement continueButton;

    @FindBy(id = "fb")
    private WebElement fbButton;

    @FindBy(id = "sign_in_button")
    private WebElement googleButton;

    public void skipToHomePage() {
        skipLink.click();
    }

    public void registerByMobileOrEmail(String mobileorEmail) {
        mobileOrEmailField.sendKeys(mobileorEmail);
        continueButton.click();
    }

    public void signInByFacebook() {
        fbButton.click();
    }

    public void signInByGoogle() {
        googleButton.click();
```

```
    }
  }
```

So, we have declared the UI elements as private and the accessor methods as public so that they can allow anyone to perform any operation on those UI elements.

These are simple, straightforward actions that a page is doing, and we believe that it belongs to that page class. There are some discussions around as to whether a Page Object should include assertions or not. We will come to this a little later; before that, let's refactor the existing code to create another page class.

Refactor-2

We will start this refactoring by creating a new package under the `java` folder of the solution called `pages`, illustrated here:

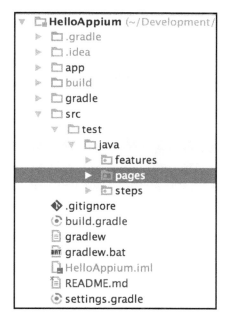

1. Right-click on the `pages`, select **New > Java Class**, and enter the name `LandingPage`
2. Copy the preceding code snippet of `LandingPage` and paste it
3. We also need to implement wait for the visibility of elements to remove any flakiness in the test

We learned about test synchronization in the last chapter, and we also learned about the importance of a proper wait strategy in the test. So, instead of having each page class create its own implementation of `WebDriverWait`, we can extract it to a base page, which can be extended by different page classes and can utilize the common methods. So, the next step is to add a `BasePage`:

1. Add another Java class to the pages package and name it as `BasePage`

2. WebDriver wait needs a driver instance and for now, we will create only one method, `waitForElementToBeVisible()`.

3. Copy this code in the `BasePage`:

```java
public class BasePage {
    private AppiumDriver driver;
    private WebDriverWait wait;

    public BasePage(AppiumDriver driver) throws Exception {
        this.driver = driver;
        wait = new WebDriverWait(this.driver, 30);
    }

    public void waitForElementToBeVisible(WebElement element) {
        wait.until(ExpectedConditions.visibilityOf(element));
    }
}
```

4. At this point, we need to make the `LandingPage` class extend the `BasePage` and add the method we just created to the `LandingPage` class method `skipToHomePage()`. This will make you add the constructor matching the `superclass`; go ahead and do that. Here's how it will look after the changes:

```java
public void skipToHomePage() {
    waitForElementToBeVisible(skipLink);
    skipLink.click();
}
```

At this point, we can try running the code but it will throw a `NullPointerException`. PageFactory supports this pattern and helps cut down the code. Let's see what changes we need to make and where.

The reason the code threw up `NullPointerException` is because the fields are not instantiated; hence, we need to initialize the PageObject. Take a look at this code:

```
skipLink.click();
```

When `PageFactory` is initialized, it is equivalent to the following:

```
appiumDriver.findElement(By.id("skip")).click();
```

Now let's try to stabilize the code here. We have added a `BasePage` class, as shown just now. Let's navigate back to the `LandingPage` class; we need to add this `PageFactory` initialization:

```
PageFactory.initElements(appiumDriver,this);
```

We need to add a constructor that takes care of the `PageFactory` initialization. The code after extending `BasePage` class and adding initialization is as shown:

```
public class LandingPage extends BasePage {
    AppiumDriver appiumDriver;

    @FindBy(id = "skip")
    private WebElement skipLink;

    @FindBy(id = "login_register_view")
    private WebElement mobileOrEmailField;

    @FindBy(id = "continue_login")
    private WebElement continueButton;

    @FindBy(id = "fb")
    private WebElement fbButton;

    @FindBy(id = "sign_in_button")
    private WebElement googleButton;

    public LandingPage(AppiumDriver appiumDriver) throws Exception {
        super(appiumDriver);
        this.appiumDriver = appiumDriver;
        PageFactory.initElements(appiumDriver, this);
    }

    public void skipToHomePage() {
        waitForElementToBeVisible(skipLink);
        skipLink.click();
    }

    public void registerByMobileOrEmail(String mobileorEmail) {
```

```
        mobileOrEmailField.sendKeys(mobileorEmail);
        continueButton.click();
    }

    public void signInByFacebook() {
        fbButton.click();
    }

    public void signInByGoogle() {
        googleButton.click();
    }

}
```

To use the preceding code, we need to change the `public void iChooseAsMyCity(String city)` method and edit a couple of lines. Consider the following code:

```
Wait wait = new FluentWait(appiumDriver)
        .withTimeout(10, TimeUnit.SECONDS)
        .pollingEvery(250, TimeUnit.MILLISECONDS)
        .ignoring(NoSuchElementException.class)
        .ignoring(TimeoutException.class);
wait.until(ExpectedConditions.visibilityOfElementLocated(By.id("skip")));

appiumDriver.findElement(By.id("skip")).click();
```

We can substitute it with the following line and run the same test again. It will pass seamlessly and look at the readability and code organization we have introduced:

```
new LandingPage(appiumDriver).skipToHomePage();
```

On similar lines, we can create a `HomePage` class that will have elements such as select city dropdown. Let's do the exercise of creating the `HomePage` class. In the `HomePage` class, we need not map everything that is present on the UI; it can just be those elements that you need interaction with. Here's the implementation of the same:

```
public class HomePage extends BasePage {
    AppiumDriver appiumDriver;

    @FindBy(id = "citySpinner")
    private WebElement cityDropdown;

    @FindBy(id = "search_ET")
    private WebElement citySearchBox;

    @FindBy(id = "city_name")
```

```
    private WebElement cityName;

    @FindBy(xpath = "//android.widget.TextView[@text='Cars']")
    private WebElement mobileOrEmailField;

    @FindBy(id = "sign_in_button")
    private WebElement googleButton;

    public HomePage(AppiumDriver appiumDriver) throws Exception {
        super(appiumDriver);
        this.appiumDriver = appiumDriver;
        PageFactory.initElements(appiumDriver, this);
    }

    public void selectCity(String city) {
        cityDropdown.click();
        citySearchBox.click();
        citySearchBox.sendKeys(city);
        waitForElementToBeVisible(cityName);
        cityName.click();
    }
}
```

With the preceding piece of code, we can replace the following line:

```
appiumDriver.findElement(By.id("citySpinner")).click();
appiumDriver.findElement(By.id("search_ET")).click();
appiumDriver.findElement(By.id("search_ET")).sendKeys(city);

wait.until(ExpectedConditions.visibilityOfElementLocated(By.id("city_name")
));
appiumDriver.findElement(By.id("city_name")).click();
```

This is the line it will be replaced with:

```
new HomePage(appiumDriver).selectCity(city);
```

So, the iChooseAsMyCity(String city) method now looks like this:

```
@And("^I choose \"([^\"]*)\" as my city$")
public void iChooseAsMyCity(String city) throws Throwable {
    new LandingPage(appiumDriver).skipToHomePage();
    try {
        if
(appiumDriver.findElement(By.xpath("//android.widget.Button[@text='Later']"
)).isDisplayed())
appiumDriver.findElement(By.xpath("//android.widget.Button[@text='Later']")
).click();
```

```
    } catch (Exception e) {
        //Do nothing
    }
    new HomePage(appiumDriver).selectCity(city);
}
```

This code is much more readable and easy to maintain. We can even add this handling of pop up message to upgrade by clicking on **Later** to HomePage or BasePage itself so that the code will look much cleaner. Now that we have seen how to do this, I am leaving it to you to implement the same. Below is the pictorial difference in the code readability and structure:

The next exercise for you is to replace the existing code we have written and model it all on the Page Object concept.

Assertions

Assertions are the core of test automation, and there has been a good long debate on where assertions belong. Broadly, there are two types of approaches for handling assertions, and they can be implemented in either of the following:

- Page Object
- Test script

The first approach says that Page Objects should contain assertions. The advantage of this approach is to minimize the duplication of assertions in the test suite. Also, it helps in organizing the messages and following the *Tell, Don't Ask* principle (for more information visit: https://martinfowler.com/bliki/TellDontAsk.html). The *Tell, Don't Ask* principle recommends that an object can be issued a command to perform some operation or logic, rather than to query its state. It suggests that we should tell the object what to do, rather than asking the object for data and then acting on it.

Let's apply the same in our code. Here's what we have automated:

```
@Then("^I should see the first car search result with \"([^\"]*)\"$")
public void iShouldSeeTheFirstCarSearchResultWith(String arg0) throws
Throwable {
    List<WebElement> elements =
appiumDriver.findElements(By.id("cars_ad_list_title_tv"));
    Assert.assertTrue(elements.get(0).getText().startsWith(arg0));
}
```

Implementing assertions in Page Object

The first step is to replace the implementation with the following page class and the method that will perform the verification of the search results on the car search results page. So, we can declare something like this:

```
@Then("^I should see the first car search result with \"([^\"]*)\"$")
public void iShouldSeeTheFirstCarSearchResultWith(String arg0) throws
Throwable {
    new CarResultsPage(appiumDriver).verifySearchResult(searchInput);
}
```

The next step is to implement the `CarResultsPage` class and the `verifySearchResult(String searchInput)` method. So, this method will take care of the verification of the result. This is the implementation of the `CarResultsPage` class:

```java
package pages;

import io.appium.java_client.AppiumDriver;
import org.junit.Assert;
import org.openqa.selenium.WebElement;
import org.openqa.selenium.support.FindBy;
import org.openqa.selenium.support.PageFactory;

import java.util.List;

public class CarResultsPage extends BasePage {
    AppiumDriver appiumDriver;

    @FindBy(id = "category")
    private WebElement categoryChooser;

    @FindBy(id = "inspected_checkbox")
    private WebElement inspectedToggle;

    @FindBy(xpath = "//android.widget.TextView[@text='SORT']")
    private WebElement sortLink;

    @FindBy(xpath = "//android.widget.TextView[@text='FILTER']")
    private WebElement filterLink;

    @FindBy(id = "cars_ad_list_title_tv")
    private List<WebElement> searchResultText;

    public CarResultsPage(AppiumDriver appiumDriver) throws Exception {
        super(appiumDriver);
        this.appiumDriver = appiumDriver;
        PageFactory.initElements(appiumDriver, this);
    }

    public void verifySearchResult(String text) {
        for (WebElement result : searchResultText) {
            Assert.assertTrue(result.getText().contains(text));
        }
    }

}
```

we can add other assertions and remove the assertion from the step class. Let's look at the other approach of implementing assertions in the test script.

Implementing assertion in test script

The second approach is having assertions in the test suite. In this case, we will have a Page Object that implements a getter for the element state we want to have a check on or verify upon. So in this case, we need a method in the page class that will return us the text of the search result header.

Let's implement the page class for this:

```
package pages;

import io.appium.java_client.AppiumDriver;
import org.openqa.selenium.WebElement;
import org.openqa.selenium.support.FindBy;
import org.openqa.selenium.support.PageFactory;

import java.util.List;

public class CarResultsPage extends BasePage {
    AppiumDriver appiumDriver;

    @FindBy(id = "category")
    private WebElement categoryChooser;

    @FindBy(id = "inspected_checkbox")
    private WebElement inspectedToggle;

    @FindBy(xpath = "//android.widget.TextView[@text='SORT']")
    private WebElement sortLink;

    @FindBy(xpath = "//android.widget.TextView[@text='FILTER']")
    private WebElement filterLink;

    @FindBy(id = "cars_ad_list_title_tv")
    private List<WebElement> searchResultText;

    public CarResultsPage(AppiumDriver appiumDriver) throws Exception {
        super(appiumDriver);
        this.appiumDriver = appiumDriver;
        PageFactory.initElements(appiumDriver, this);
    }

    public String getFirstSearchResult() {
```

```
        return searchResultText.get(0).getText();
    }

}
```

Now, let's go to the step implementation and refactor a couple of things there. First, we need to call the preceding method, store the result, and then make the necessary assertions:

```
@Then("^I should see the first car search result with \"([^\"]*)\"$")
public void iShouldSeeTheFirstCarSearchResultWith(String searchInput)
throws Throwable {
    String searchResult = new
CarResultsPage(appiumDriver).getFirstSearchResult();
    Assert.assertTrue(searchResult.startsWith(searchInput));
}
```

Now, we can execute the test to get the same result. Let's discuss some other practices of test development that will help us create a better test automation framework, and it applies to mobile test automation solutions as well.

Avoiding dependencies between tests

Each test we author should be independent of the others. Developers or testers using the solution should be able to run any test in any order based on the need. Generally, when we submit the cucumber feature to run, scenarios need not execute in the same order and, as a result, the test will easily be broken if there are dependencies in it. Hence, it becomes easy when we execute via cucumber as we follow the Given-When-Then format.

Introducing set up and tear down

Most of the tests that we write can be broken into three parts:

- Pre-condition
- Action and verification
- Post-condition

Pre-condition takes the app under test to a certain desired state. In our case, it will translate to install the app on the device, log in to the app, and come on the respective screen.

The action will translate to tapping on car category, and searching for a car. Verification will translate to asserting if we have the correct result, as expected.

Post-condition will translate to logging out of the app, cleaning of the app state, and even uninstalling the app.

Cucumber exposes two hooks that take care of running pre-condition and post-condition using the @Before and @After hooks. These hooks are very similar to the setup and teardown methods provided in **xUnit** testing tools. Both before and after are global hooks; hence, they can be declared in any step.

@Before allows you to run a block of code before every scenario. We can declare many methods tagged with the @Before hook. They run in the same order as they are declared. In our code base, we have declared a method with the @Before tag. So, it is executed before running any scenario:

```
@Before
public void startAppiumServer() throws IOException {
}
```

@After allows you to run a block of code after the last step of each scenario. It runs regardless of the status of the last step, be it failing, undefined, skipped, or pending. It runs in the opposite order of declaration. We have declared a method with the @After tag in our code base. Hence, it takes care of the session and what to do after running the test:

```
@After
public void closeAppiumServerSession() {
}
```

This finishes the framework designing principles and some of the concepts.

Summary

In this chapter, we learned about the Page Object design pattern and how it can be used to give a structure to the code we have written. We also went through refactoring, understanding the design pattern and how it has significantly improved the code readability and makes the maintenance look easier. We learned about assertions and how they can be used. We also learned about where assertion belongs and the pros and cons of each approach. We discussed some framework design principles of avoiding the dependent test designs and the importance of hooks, such as @Before and @After, provided by cucumber.

Now we have a decent framework in structure and the tests are a little mature with the design pattern in place. The next step is to be able to run the test on different targets, such as an emulator and an actual device, understand the hassles around it, and solve them.

9
How to Run Appium Test on Devices and Emulators

In the last chapter, we were exposed to design patterns, and we learned how to structure code for better readability and maintenance. We have a decent test that deploys an app on the target device, launches the app, and performs a search. The next stage in Appium is to be able to run these tests on an emulator and actual device. In this chapter, we will study the following topics in detail:

- Emulator:
  - Setting up and configuring
  - Running the test on the emulator
- Devices:
  - How to configure
  - Running the test on devices

Emulator

An emulator is an application that emulates a real mobile device, which lets you prototype the app under development or allows you to test out the app without actually buying a physical device. When we install Android SDK, we can create emulators based on the available API level, CPU, and RAM. We learned how to set up an Android Virtual Device using Android SDK in `Chapter 2`, *Setting Up the Machine*. We also briefly learned about the emulator and how to download one virtual device.

In this chapter, let's take a detailed look into Genymotion, which provides Android emulators that are faster and better performing compared to Android SDK:

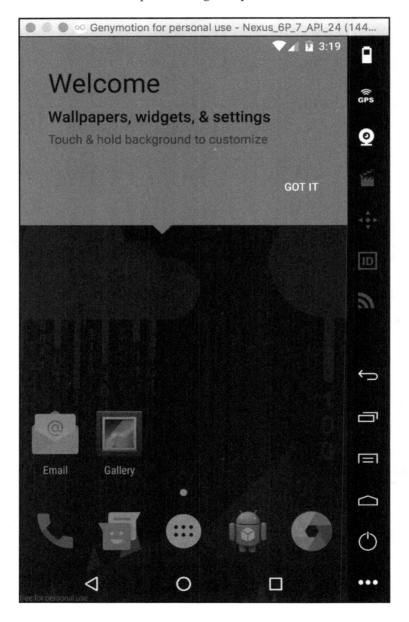

To install app on the Genymotion emulator, the normal `adb` commands will work fine, as shown:

```
adb install /path/to/app/<app_name>.apk
```

If the app under test is dependent on Google Play, we need to perform these steps:

1. Look for the Google Play Store APK, `com.android.vending-x.x.xx.apk`, for the device API level and install it.
2. Flash the emulator with the respective Google play Service's `gapps-lp-YYYYMMDD-signed` file.
3. Restart the emulator.
4. Launch the Play Store and update the google apps installed.

The advantages of using the GenyMotion emulator are:

- The Genymotion emulator is a better performant than the Android SDK emulators. Genymotion uses the x86 architecture to run the Android virtualization.
- Genymotion emulators don't crash as frequently compared to Android Virtual Devices.
- Genymotion has a larger array of devices to create an emulator from.

Running test on the Genymotion emulator

When you are running a single Genymotion emulator on your machine, you can pass the platform version to the desired capabilities and it will take care. The code for that is as given; customize it for the platform version you have created the emulator for:

```
capabilities.setCapability("platformVersion", "5.1");
```

When we are running multiple versions of Genymotion emulator, we need to pass the `udid` of the targeted device where we want to run the automation:

```
[➜  ~ adb devices
List of devices attached
192.168.56.102:5555       device
192.168.56.101:5555       device
```

Here's the code snippet for passing the `udid` as the desired capabilities:

```
capabilities.setCapability("udid", "192.168.56.102:5555");
```

With the Genymotion emulator, one of the exceptions is that the desired capability to launch the emulator doesn't work the way it works seamlessly in Android Virtual Devices. Also, we need to launch Genymotion emulators before we trigger the test. Here, how we can launch the Genymotion emulator via the command line is explained.

Use these steps to start the Genymotion emulator via the command line on macOS:

1. Launch the Terminal.
2. Type in the `VBoxManage list vms` command. This is the sample output:

```
[➜  ~ VBoxManage list vms
"Nexus_5_API_21_B" {5b824ed3-a8bc-4662-b623-ef33701ccf91}
"Nexus_6P_7_API_24" {9aacea6a-0647-4471-b57c-86617989297a}
"Nexus_6P_7_API_24_Clone" {1010b597-fb0b-4ef7-9f59-57070b3108a7}
```

3. Type in the following command (modify the vm name with the data on your system):

```
open -a /Applications/Genymotion.app/Contents/MacOS/player.app --
args --vm-name '1010b597-fb0b-4ef7-9f59-57070b3108a7'
```

4. The last parameter passed is `vm id`, as shown in the .

This will launch the Genymotion emulator without launching the Genymotion app. Let's take a look at how to run the same test for physical devices.

Devices

To do any development and debugging activity on Android devices, the first thing we need to do is enable the developer options. Different phones have different navigations for enabling developer options; here, we list a few of them:

- Samsung Phones:
    - Launch **Settings** > **About Device** > **Build number**
- LG Phones:
    - Launch **Settings** > **About Phone** > **Software Information** > **Build number**
- Stock Android Phone:
    - Launch **Settings** > **About phone** > **Build number**:

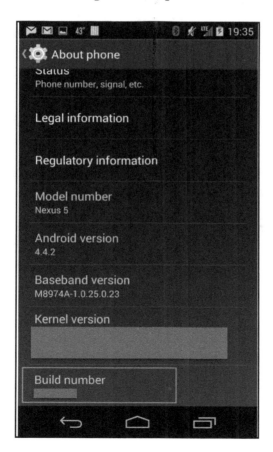

Once we reach the **Build number**, we need to tap on it seven times, and then it will show a message saying **You are now a developer!**. This will enable the developer options on the device under the **Settings** menu. Tap on **Developer options** and select **USB debugging**. Also, ensure that the option of **Verify apps over USB** is turned off. This option, when turned on, stops app deployment on the physical device:

This will show a popup (as illustrated), on which we need to press **OK**:

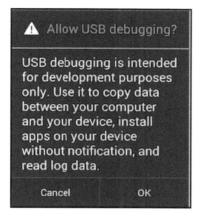

Once the preceding setups are done on the device, we can connect the device to the machine, launch the Terminal (the Command Prompt in the case of Windows), and type in this command:

```
adb devices
```

The expected output is shown here; we have one physical device running with UDID 2a2d916 and one GenyMotion emulator with ID 192.168.56.101:5555:

Sometimes, when the devices still don't show up in the output, we can run through the following steps to fix this (these are macOS-specific steps):

1. Open the USB manager on your machine (macOS).
2. Use the **vendor ID** (highlighted in red in the following screenshot) and update it in the adb_usb.ini file.

To obtain the **vendor ID** on macOS:

1. Click on the Apple icon in the top-left of the screen.
2. Click on **About This Mac**.
3. On the popup, tap on **System Report**.

4. Under the hardware section, click on **USB**.

5. You will notice the device connected there; click on the **Android** device:

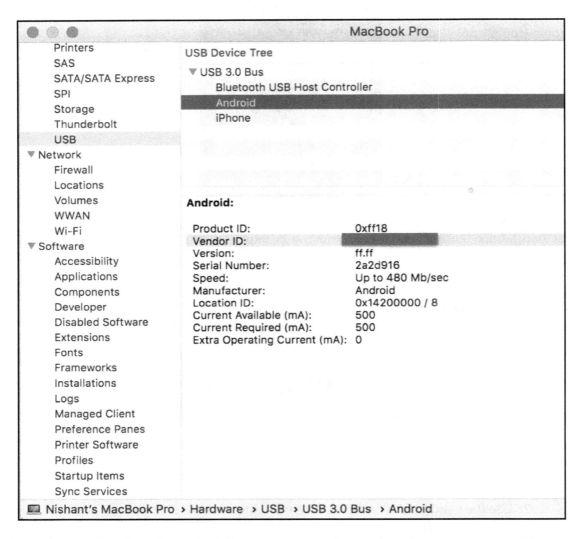

Copy the **Vendor ID** and run the following command to update the `adb_usb.ini` file:

```
vim ~/.android/adb_usb.ini
```

The preceding command will open the file in edit mode, and we can paste the vendor ID in the new line there, save the file, and quit. Here's a sample snapshot of the file:

Once we are done editing this file, we can restart the adb server and run the `adb devices` command, as shown:

```
adb kill-server
adb start-server
adb devices
```

Running a test on actual Android devices

Running a test on a physical device is very similar to running the test on emulators once the preceding setup is done. As long as we are providing the platform version correctly and providing there is only a single device connected, the test will pick up that device for execution. However, when we have multiple devices connected with the same android platform version, we need to specify the unique udid of the device to run the test.
So, if we have the emulator and device of the same platform version, we can use the following line and update the respective device ID to run the test on the connected physical device:

```
capabilities.setCapability("udid", "2a2d916");
```

Running a test on actual iOS devices

Until now, we have largely built the code base for Android, but most of it remains the same for iOS. We need to create a similar project for iOS apps and use an `.app` or `.ipa` file to deploy the app on device/simulator. The current project we have developed is Android-specific; however, we can reuse the feature file. Locators might be different in the case of the iOS app, but the steps largely remains the same to obtain the locator.

However, when it comes to running tests on iOS devices, we need to go through the following series of steps before triggering the test.

The first major requirement is to have a macOS as your machine and, second, to have the iOS app under test signed with a development provisioning profile.
If we are using a physical device, we need to enable **UI automation** in the developer options in the iPhone device. These are the steps to do this:

1. Switch off the iDevice.
2. Connect it to the Mac running Xcode.
3. Switch it back on to have **Developer** options appear under device **Settings**.
4. Tap on the **Developer** option.
5. **Enable UI automation**.

Now, the device is ready to run any Appium tests. The next step to run the Appium test is to get the UDID of the devices. The following steps help you obtain the UDID of iOS devices:

1. Connect your iOS Device to your Mac and launch **iTunes**.
2. In the left pane, go to **Devices** > **Select your Device**.
3. In the right pane, reveal the identifier by clicking on **Serial Number**. It's a clickable element that toggles.
4. Copy the device identifier and save it.

We need to have a couple of libraries installed before we run the test. Run the following commands to install these libraries:

* `ios-webkit-debug-proxy`: It proxies requests from the `usbmuxd` daemon over a websocket connection. It allows you to send commands to MobileSafari and UIWebViews on real and simulated iOS devices:

```
brew install ios-webkit-debug-proxy
```

- `libmobiledevice`: It's a cross-platform **software library** that talks the protocols to support Apple devices. It supports iOS devices natively:

```
brew install libimobiledevice
```

- `usbmuxd`: It stands for USB multiplexing daemon and is in charge of multiplexing connections over USB to an iOS device:

```
brew install usbmuxd
```

- `carthage`: It's a dependency manager for a Cocoa application. Appium uses the Facebook `WebDriver` agent, which in turn needs `carthage` as a dependency manager:

```
brew install carthage
```

- `ios-deploy`: `ideviceinstaller` doesn't work with iOS 10 yet. Hence, we need to use the `ios-deploy` library to interact with real devices. Use this line to implement it:

```
npm install -g ios-deploy
```

These are the steps to follow before triggering a test on an iOS physical device:

1. Launch the **Terminal**.
2. Run `ios-webkit-debug-proxy` by running the following command. This command restricts the proxy to just one device identified by its UDID:

```
ios_webkit_debug_proxy -c <UDID>:27753
```

3. Assuming that the test is starting the Appium server, run the Appium test. A sample of the desired capabilities will look like this:

```
capabilities.setCapability("platformName", "iOS");
capabilities.setCapability("platformVersion", "9.3");
capabilities.setCapability("deviceName", "iPhone");
capabilities.setCapability("udid",
"2b6f0cc904d137be2e1730235f5664094b831186");
```

So, the preceding steps will help run the Appium test on physical iOS devices. For running the test on physical iOS devices, `udid` is a must; we don't need the `udid` as a desired capability but the `deviceName` has to match the simulator name for iOS simulators.

Summary

In this chapter, we learned how to set up Genymotion emulators and how to configure them. We learned how to alter the desired capabilities to run the test on emulators. We learned how to set up Android devices for development and testing by turning on the developer options. We also learned how to turn on USB debugging and run the test on an Android device by passing the `udid`.

We explored different libraries to install (via Hómebrew) for running the Appium test on an actual iOS device. We also discussed how to get the UDID of iOS devices. We went through the steps to start `ios-webkit-debug-proxy` and the desired capabilities to use for an iOS test.

In the next chapter, we will learn how to run the Appium test via the continuous integration tool, Jenkins. We will go through the detailed process of setting up Jenkins and running the test.

10
Continuous Integration with Jenkins

In the last chapter, we looked at how to run the Appium test on an emulator and physical devices. We also learned how to start the emulator through the command line. We explored how to run the Appium test on physical devices, including iOS devices. So far, we have seen how to use Appium, learned how to author test, learned to automate gestures, and learned about design patterns as well. The next step is to run these Appium tests via a continuous integration tool, Jenkins. In this chapter, we will take a detailed look at the following:

- Setting up Jenkins
- Exporting reports as artefacts

Generally, on any development project, we use a continuous integration tool. It's a standard development practice that requires developers to integrate code into a shared repository. Once the developer checks in the code, it is verified by the automated build that does basic jobs, such as compiling the code and running unit tests.

Before we set up Jenkins, let's refactor the code to run the automation test via command line using the tool `Gradle`.

Refactoring -1

Until now, we have been running the test via an IDE. When we started with the `gradle` file, it was majorly to pull in the dependencies needed for the project. Here's how the current version of the `gradle` file looks:

```
group 'com.test'
version '1.0-SNAPSHOT'

apply plugin: 'java'

//sourceCompatibility = 1.8

repositories {
    mavenCentral()
}

dependencies {
    testCompile group: 'junit', name: 'junit', version: '4.11'
    compile group: 'info.cukes', name: 'cucumber-java',
    version: '1.2.5'
    compile group: 'io.appium', name: 'java-client',
    version: '5.0.0-BETA6'
}
```

The next step is to create a task that will execute the cucumber features in a different feature file. A task represents an atomic piece of work for a build. Tasks generally belong to a project and the syntax to define a task is this:

```
task taskName (type: someType){
configuration
}
```

A task is made up of a sequence of actions; some typical actions can be added by calling `doFirst()` or `doLast()`. So, let's go ahead and add a task to execute all features and generate a `.json` report. Copy the following code snippet and paste it in the `gradle` file below the dependencies section:

```
task runAllTest(type: Test, dependsOn: ['clean', 'build']) {
    doLast {
        String tags = getTags()
        javaexec {
            main = "cucumber.api.cli.Main"
            classpath = sourceSets.test.runtimeClasspath
            args = ["-p", "pretty", "-p",
            "json:${reporting.baseDir}/cucumber/cucumber.json",
```

```
            "--glue", "steps", "-t", tags,
            "${project.projectDir}/src/test/java/features"]
        }
    }
}

private String getTags() {
    def tags = System.getProperty("tags")
    if (tags != null)
        return tags;
    return "~wip"
}
```

Let's understand the above piece of code we have written; we are essentially doing the following things:

- Creating a task called `runAllTest`
- Making it dependent on the other task `clean` and `build` (which are predefined)
- Getting a `string` (tag name) via system parameters and performing a null check on the same
- Invoking cucumber CLI with a bunch of arguments:
    - `-p`: To create a pretty report in Json format in the specified directory
    - `--glue`: To find the step implementation in the "steps" package
    - `-t`: To filter the features file (in the specified path) based on the passed string, which is tags

So, the entire `gradle` file should look as illustrated:

```
group 'com.test'
version '1.0-SNAPSHOT'

apply plugin: 'java'

//sourceCompatibility = 1.8

repositories {
    mavenCentral()
}

dependencies {
    testCompile group: 'junit', name: 'junit', version: '4.11'
    compile group: 'info.cukes', name: 'cucumber-java', version: '1.2.5'
    compile group: 'io.appium', name: 'java-client', version: '5.0.0-BETA6'
}

task runAllTest(type: Test, dependsOn: ['clean', 'build']) {
    doLast {
        String tags = getTags()
        javaexec {
            main = "cucumber.api.cli.Main"
            classpath = sourceSets.test.runtimeClasspath
            args = ["-p", "pretty", "-p", "json:${reporting.baseDir}/cucumber/cucumber.json",
                    "--glue", "steps", "-t", tags, "${project.projectDir}/src/test/java/features"]
        }
    }
}

private String getTags() {
    def tags = System.getProperty("tags")
    if (tags != null)
        return tags;
    return "~wip"
}
```

If you note the preceding code, we are accepting tags as an input to run the test. So, let's add a tag on the scenario and call it @search:

```
@search
Scenario: Search for a used Honda City car in Bangalore city

When I launch Quikr app
And I choose "Bangalore" as my city
And I search for "Honda City" under Used Cars
Then I should see the first car search result with "Honda"
```

Now we can pass these created tags using the command line; let's test the preceding gradle task by following these steps:

1. Launch Emulator or connect a device.
2. Change the desired capability to match the PlatformVersion of the emulator/device.
3. Launch the Terminal (the **Command Prompt** on Windows) and navigate to the project root folder:

```
[-> HelloAppium pwd
/Users/nishant/Development/HelloAppium
 -> HelloAppium gradle clean build runAllTest -Dtags=@search
```

4. Type in this command and press *Enter*:

```
gradle clean build runAllTest -Dtags=@search
```

Windows user can run the command

```
./gradlew clean build runAllTest -Dtags=@search
```

This should start the test on the targeted device. We can add the same tag to other test as well and run them.

Once we have this up-and-running, we are good to set up Jenkins.

Setting up Jenkins

Jenkins is an open source continuous integration tool that helps in automating development-related repetitive tasks. It runs as a local server on a host machine where we install it:

Let's follow these steps to install Jenkins:

1. Download the Jenkins mac OS X installer or Windows installer from `http://jenkins-ci.org`.
2. Double-click on the `.pkg` (`.msi` for Windows) file to install Jenkins and select the location installation.
3. Once it is successfully installed, the browser will open to `http://localhost:8080`.
4. The browser will redirect to `http://localhost:8080/login?from=%2F` with a message for macOS X and Windows.

Unlock Jenkins

To ensure that Jenkins is securely set up by the administrator, a password has been written to the log (not sure where to find it?) and this file on the server:

`/Users/Shared/Jenkins/Home/secrets/initialAdminPassword`. Copy the password from either location and paste it below as shown in the following screenshot:

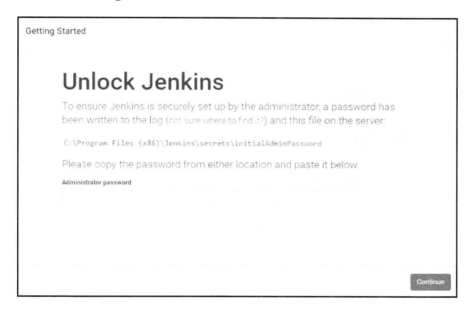

5. Use the following command to view the password and copy it:

   - For macOSX : `sudo cat /Users/Shared/Jenkins/Home/secrets/initialAdminPassword`

   - For Windows: Navigate to the earlier mentioned location (in the Getting Started pop up) and open the file with **Notepad**; copy the password.

6. Enter the password in the Jenkins log-in page, and it will show you the screen to install and manage the plugin.
   1. Close that and click on **Start Jenkins.**

This completes the Jenkins setup on your machine. The next step is to create a job that runs the automation suite, but we need to implement the version control system (Git) with our current project before that.

Moving a project to Git

Until now, whatever we have coded resides locally on our machine, which will never be an ideal case as we will typically be using the source control tool GitHub, Bitbucket, and so on. Follow the given steps to move the project to GitHub (assuming that you have a GitHub account; if not, please sign up on https://github.com/):

1. Install Git by downloading the respective installer for your machine (either Mac or Windows).

2. Once done, launch the Terminal and type in the `git --version` command. It should show something similar to this (with a higher version number):

```
[➜  ~ git --version
git version 2.6.4
➜  ~ ▊
```

3. The next step is to configure your git username and email using the following commands:

```
$ git config --global user.name "firstname lastname"
$ git config --global user.email "firstname.lastname@xyz.com"
```

Once the preceding steps are executed, we are done with the setup of Git; the next step is to move the repository to Git. Follow these steps for that:

1. Log in to Git and click on **New repository**. You will be directed to another window, as shown:

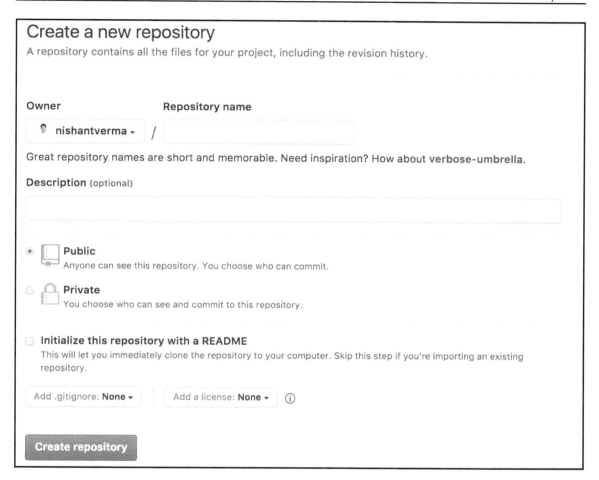

2. Enter a **Repository Name**.
3. Enter the **Description** for your repository.
4. By default, **Public repository** will be selected. Choose **Private** if you want to set up a private repo.
5. Click on **Create repository**.

Once done you will see this screen:

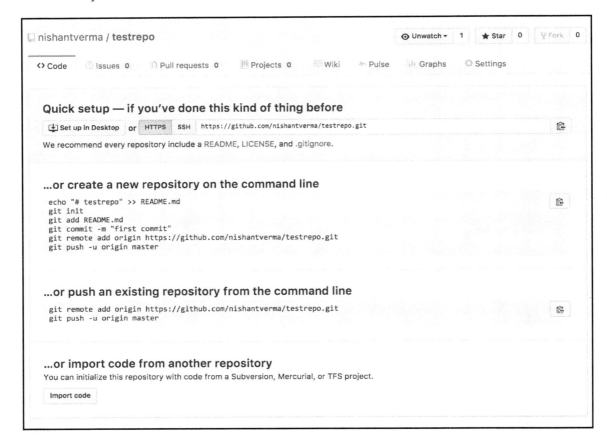

So, now we have an existing project that we want to get started with. Let's follow these steps to push the HelloAppium project to GitHub:

1. Launch the Terminal (the Command Prompt on Windows) and navigate to the HelloAppium project folder.
2. Type git init, and you will see this output:

```
[→ HelloAppium git init
Initialized empty Git repository in /Users/nishant/Development/HelloAppium/.git/
→ HelloAppium git:(master) ✗ ▊
```

3. Once done, run the git remote add origin https://github.com/nishantverma/HelloAppium.git command to add the remote origin.
4. Do a git add for all files, as shown:

   git add --all

5. Do a commit with the "Initial Commit" message:

   git commit —m "Initial Commit"

6. Once done, it will show a bunch of files that are ready to be pushed.
7. Run the push command:

   git push —u origin master

8. This is a snapshot of all the commands and the typical output:

```
[→ HelloAppium git init
Initialized empty Git repository in /Users/nishant/Development/HelloAppium/.git/
[→ HelloAppium git:(master) ✗ git remote add origin https://github.com/nishantverma/HelloAppium.git
[→ HelloAppium git:(master) ✗ git add --all
[→ HelloAppium git:(master) ✗ git commit -m "Initial Commit"
[master (root-commit) 0a44b28] Initial Commit
 19 files changed, 833 insertions(+)
 create mode 100644 .gitignore
 create mode 100644 README.md
 create mode 100644 app/old-quikr.apk
 create mode 100644 app/quikr.apk
 create mode 100644 build.gradle
 create mode 100644 gradle/wrapper/gradle-wrapper.jar
 create mode 100644 gradle/wrapper/gradle-wrapper.properties
 create mode 100755 gradlew
 create mode 100644 gradlew.bat
 create mode 100644 settings.gradle
 create mode 100644 src/test/java/features/Sample.feature
 create mode 100644 src/test/java/pages/BasePage.java
 create mode 100644 src/test/java/pages/CarResultsPage.java
 create mode 100644 src/test/java/pages/HomePage.java
 create mode 100644 src/test/java/pages/LandingPage.java
 create mode 100644 src/test/java/steps/HomePageSteps.java
 create mode 100644 src/test/java/steps/HomePageStepsWithWait.java
 create mode 100644 src/test/java/steps/HomePageWebSteps.java
 create mode 100644 src/test/java/steps/StartingSteps.java
[→ HelloAppium git:(master) git push -u origin master
Counting objects: 30, done.
Delta compression using up to 4 threads.
Compressing objects: 100% (25/25), done.
Writing objects: 100% (30/30), 30.08 MiB | 79.00 KiB/s, done.
Total 30 (delta 2), reused 0 (delta 0)
remote: Resolving deltas: 100% (2/2), done.
To https://github.com/nishantverma/HelloAppium.git
 * [new branch]      master -> master
Branch master set up to track remote branch master from origin.
→ HelloAppium git:(master)
```

We have pushed our project to GitHub with the preceding steps. The next job is to create the Jenkins task that will use this repo to run the Appium test.

Adding Jenkins plugin

To start the Jenkins setup, we need to install a couple of plugins that Jenkins provides to make the process easier. We need a couple of Jenkins plugins to help set up the automated test run. Follow these steps to install some of the plugins:

1. Launch Jenkins (`http://localhost:8080`).
2. Click on **Manage Jenkins**.

3. Select **Manage Plugins**, as illustrated:

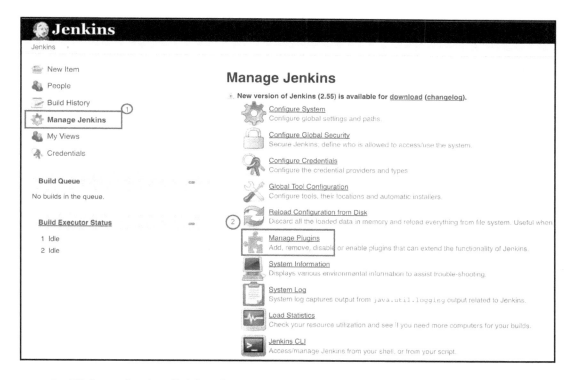

4. Click on the **Available** tab.
5. Click on **Filter** and type in **Gradle plugin**.
6. Select the **checkbox** next to the result.
7. Click on **Install without restart**.
8. Repeat the preceding steps for the following plugins:
   1. **Cucumber reports**
   2. **Github plugin**
   3. **Android Emulator plugin**
9. Once done, restart Jenkins.

This finishes the installation of all the required plugins point in time. We can always go ahead and add more plugins as the need arises.

Setting up the Jenkins task

Once the preceding plugins are installed, it becomes slightly easier for us to use these plugins to set up the Jenkins task. Follow the given steps to create the Jenkins task:

1. Launch Jenkins (`http://localhost:8080`).

2. Click on **Manage Jenkins** > **Configure System**.

3. Under **Global properties**, select **Environment variables** and add the **Name** and **Value**. **Value** should be local to your machine; it should be what we have set up in the bash profile in `Chapter 2`, *Setting Up the Machine*:

```
Name : ANDROID_HOME
Value : /usr/local/Cellar/android-sdk/24.4.1_1
```

4. Click on **Save.**

5. Once done, click on **New Item**.

6. Enter a project name and select **Freestyle project**; click on **OK**:

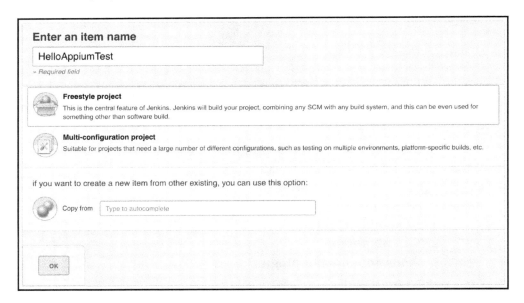

7. Enter the **Description** you want for this project.

8. Under **Source Code Management**, select **Git**.
9. Enter the **Repository URL** and add **Credentials** if it's a private repository, as illustrated:

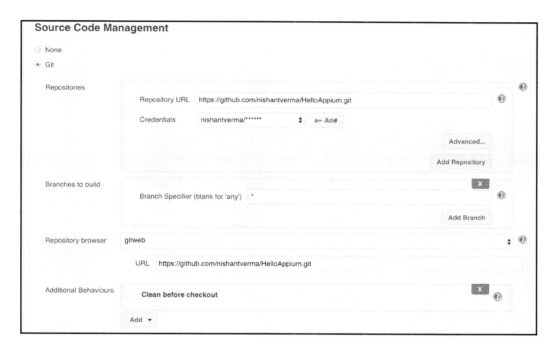

10. Choose `gitweb` as the **Repository browser**.
11. Enter the **URL** again for the **Repository browser**.
12. Under **Additional Behaviours**, click on **Add** and select **Clean before checkout**.
13. For now, we will manually trigger the builds; hence, we need not select any of the options under **Build Triggers**.
14. Click on **Add build step** under **Build** and select **Execute shell**. Enter this command:

```
./gradlew clean build runAllTest -Dtags=@search
```

In this **Command** textbox as shown:

15. Alternatively, we can select **Invoke Gradle script**, which will allow us to use the Gradle task directly. Refer to the following screenshot. For now, choose between above mentioned points 14 or 15:

16. Click on **Add post-build action**.

17. Select **Cucumber reports** from the drop-down, as follows:

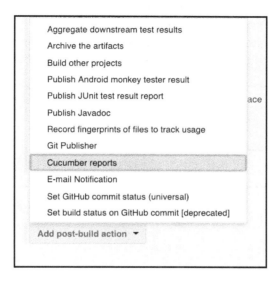

18. Click on **Save.**
19. We will see this screen once the task is created:

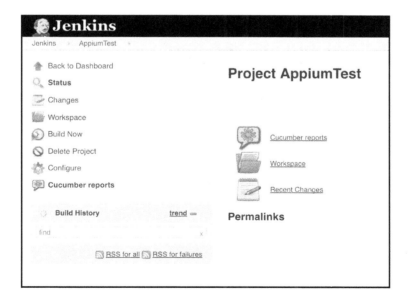

20. Launch the **Genymotion emulator**. We haven't added a step to do this in an automated way. So for now, we will have to start it manually.
21. Click on **Build Now** in the left panel.
22. Under **Build History**, click on the running job drop-down and select **Console Output**. This will show the runtime log of the running job:

We have finished setting up the Jenkins task and we have also seen how to run the task. The next step is to view the reports. This section will help you understand this.

Viewing reports in Jenkins

Once the preceding job is complete, Jenkins will show you some of the information it collects as part of the result. Refer to the following screenshot; it shows **Cucumber reports**, **Workspace,** and **Recent Changes**:

Project AppiumTest

 Cucumber reports

 Workspace

 Recent Changes

Permalinks

- Last build (#1), 4 min 34 sec ago
- Last failed build (#1), 4 min 34 sec ago
- Last unsuccessful build (#1), 4 min 34 sec ago
- Last completed build (#1), 4 min 34 sec ago

- Clicking on **Cucumber reports** will show you the summary of the test result.
- Under **Permalinks**, we can see the test result by **Last build**, **Last failed build**, and so on.
- **Recent Changes** will show the code changes commit-wise since the last run.

We have now completed setting up Jenkins to run the appium test we have authored. We can hook this to the GitHub account, where each commit will trigger the test, or we can have a manual trigger as well.

Summary

In this chapter, we covered running the appium test via a Gradle task. We learned how to pass the tags from outside to the Gradle task. We learned about Git and how to move the current project to the GitHub repo. We learned about Jenkins and how to install plugins. We also learned how to create a Jenkins task to run the test unattended. We explored how to map the Jenkins task to use the Github project for source code management and how to pass the gradle command via a shell or via the gradle task configurator. We also discussed how to enable cucumber reports and see the console output during execution time.

This pretty much completes the appium test, right from setting it up to authoring the test and configuring Jenkins to run it. In the next chapter, we will look at some of the tips and tricks that make mobile automation a little more intelligent.

11
Appium Tips and Tricks

In the last chapter, we looked at how to set up Jenkins and have a test run in an automated way. We also learned how to put the code into GitHub and then configure the Jenkins task for the purpose. We have almost come to the end of this book; in this chapter, we will learn some tips and tricks of Appium and automation in general, which can help improve our test automation and make it a little more intelligent both from the system and testing points of view.

In this chapter, we will take a detailed look at the following:

- Switching between WebView and Native
- Taking screenshots
- Recording video execution
- Interaction with an other app
- Approach for running the test in parallel
- Simulating various network conditions

Switching between views - web and native

While testing an app, we often find the need to switch between the Web and native views. A typical example is the Facebook sign-in page in many apps or an intermediate payment page. In those situations, we need to change the application context to WEBVIEW or NATIVE. Use the following code snippet to switch to WebView:

```
public static void changeDriverContextToWeb(AppiumDriver driver) {
    Set<String> allContext = driver.getContextHandles();
    for (String context : allContext) {
        if (context.contains("WEBVIEW"))
            driver.context(context);
```

```
        }
    }
```

It tries to get a list of all the context handles, checks whether there is any context that contains WebView, and then the driver switches to that context.
The following code snippet switches to native on a similar logic:

```
public static void changeDriverContextToNative(AppiumDriver driver) {
    Set<String> contextNames = driver.getContextHandles();
    for (String contextName : contextNames) {
        if (contextName.contains("NATIVE"))
            driver.context(contextName);
    }
}
```

Generally, switching between a WebView and native view happens across the app on different pages, so it will make more sense to have this method created in BasePage. The advantages of this approach are as follows:

- Easy access to call from any page
- Avoid duplication of the implementation

We can use the preceding code for reference and may tweak it, if need be. The next tip is taking a screenshot of the app while under execution.

Taking screenshots

A picture speaks a thousand words, but in our case it can save a thousand seconds. It's a good practice to take an image at the point of test failure as it will help us save a lot of time, which is needed to go through the error logs. Also, sometimes images are needed as part of the test case itself. Here are two approaches:

- Embedding a snapshot at the point of failure
- Taking a screenshot and saving it for later use or reference

Embedding a snapshot in a cucumber report becomes fa+irly easy. Cucumber exposes you to the `Scenario` interface, which makes it slightly easier to query whether the scenario has failed or passed. For example, refer to the following snapshot of code; we are doing the following step by step:

- The conditional statement helps us check whether the scenario has passed or failed
- We are checking for a failure condition in respect of the scenario
- We instruct the driver instance to take a screenshot at the point of failure:

```
if (scenario.isFailed()) {
    final byte[] screenshot = driver
            .getScreenshotAs(OutputType.BYTES);
    scenario.embed(screenshot, "image/png");
}
```

We can include the preceding code in the tear-down method. So, this will keep probing the scenario and, if it fails, it will take a screenshot and embed it in cucumber reports. If we edit the current tear-down method, it will be as shown below:

```
@After
public void tearDown(Scenario scenario) {
    try {
        if (scenario.isFailed()) {
            final byte[] screenshot = appiumDriver
                    .getScreenshotAs(OutputType.BYTES);
            scenario.embed(screenshot, "image/png");
        }
        appiumService.stop();
        appiumDriver.quit();
    } catch (Exception e) {
        System.out.println("Exception while running Tear down :"
        + e.getMessage());
    }
}
```

When we embed the failure snapshot in the current test report, it becomes more informative. Here's how a sample report with image embedding will look:

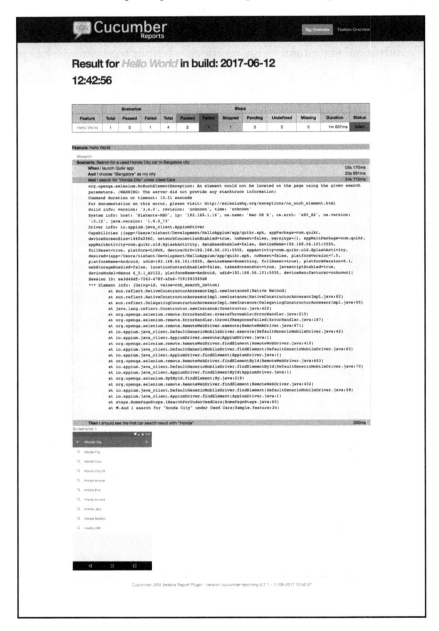

To get the above report or a nicely formatted cucumber report, we can use an external JAR listed here: `https://github.com/damianszczepanik/cucumber-sandwich`.

The second use case for taking a screenshot is to use it for manual verification. For instance, a use case would help UX team give a page by page snapshot of the app to verify the look and feel. We can use the described `getScreenshotAs()` method to taking the screenshot and store the output as a file in some predefined path. The format we are using is `.jpg`:

```
public void getScreenshot(String imageFolder) throws IOException {
    File srcImgFile=driver.getScreenshotAs(OutputType.FILE);
    String filename= UUID.randomUUID().toString();
    File targetImgFile=new File(imageFolder + filename +".jpg");
    FileUtils.copyFile(srcImgFile,targetImgFile);
}
```

Having a large number of screenshots at different points in the execution and publishing them as part of build artefacts might eat up the Jenkins agent space (assuming that the Jenkins slaves are less powerful and scaled down machine versions). We should be careful with this feature.

The next tip is to record the video execution of scenarios.

Recording video execution

Often, there is an inherent need to capture the playback when we execute a test so that we can actually see how the scenario fared. There can be a few reasons for this, one of which is the documentation. It might also be for demonstration purposes in the product team, or to see what happened on the device in the case of any failure.

Android ADB gives screen recording functionality only and not the audio capture. This should suffice for most functional test automation needs, which doesn't really require the audio component to be captured. ADB gives you a way to capture the display of Android devices, running Android 4.4 (API Level 19) or upward. The API is `adb shell screenrecord [options] <filename>`.

* Let's look at a usage example--`adb shell screenrecord /sdcard/demoVideo.mp4`:
    * The screen recording automatically stops after 3 minutes or by the `--time-limit` option, if set API usage for time limits--`adb shell screenrecord --time-limit <TIME_IN_SECONDS>`.

- The usage example for this is `adb shell screenrecord --time-limit 240`:
  - The screen record API gives the option to rotate the output by 90 degrees; however, this is just an experimental feature.
- API usage for rotate--`adb shell screenrecord --rotate`:
  - The screen record API gives the option to display log information. By default, this is off.
- API usage for displaying log info--`adb shell screenrecord --verbose`:
  - The screen record API gives the option of setting the bit rate for the video, in megabits per second. The default value is 4 Mbps. The higher the bit rate, the greater the size of video and vice versa.

API usage for bit rate--`adb shell screenrecord --bit-rate <RATE>`.

An example of this `adb shell screenrecord --bit-rate 6000000 /sdcard/demoVideo.mp4`.

A handy tip for recording video execution is to start the recording when you start the scenario; so an ideal place to call it would be in the `setup` method with the `@Before` tag. Also, adb makes only 3 minutes of screen recording; so if a scenario exceeds 3 minutes, we need to write our own logic to capture the remaining execution.

The next tip is about how we launch a different app when we have started a session with a specified app under test.

Interacting with another app

Most of the time, when we test a mobile application, it requires interaction with another app. For example, an app might need integration with the Contacts app or the SMS app. Sometimes, while testing, we might need to simulate the geo location, which can be done via an external app installed on the device/emulator (or it can even be done using Android `adb` commands).

When we start an Appium session for testing, generally it is tied to an app as we are passing the `app` parameter in the desired capabilities, so we can't really pass two apps in the desired capabilities. If we recall our code, we are using this line:

```
capabilities.setCapability("app",
"/Users/nishant/Development/HelloAppium/app/quikr.apk");
```

One way to switch between the apps is when we know the target app's **package name** and **activity name**. Android driver exposes a method, `startActivity(Activity activity)`, which basically takes an activity as input and starts it. So, a sample code snippet to start the Contacts app on a device will look like this:

```
Activity activity = new Activity("com.android.contacts",
".ContactsListActivity");
androidDriver.startActivity(activity);
```

Once we are done with the test steps we want on this app, we can use the BACK key to traverse back to the application under test:

```
androidDriver.pressKeyCode(AndroidKeyCode.BACK);
```

The `StartActivity()` method is available only for `androidDriver` and not for `AppiumDriver`. On iOS devices/simulators we can't automate two apps in one session due to a limitation from the Apple itself. The only way we can do this:

- Initiate a session 1
- Run through the steps for app 1
- Close session 1
- Start another session 2
- Run through the steps for app 2
- Close the session 2.

One thing we need to keep in mind is to set the Desired Capability `noReset` to be `true` while creating the driver instance.

Let's take a look at how we can run the test in parallel.

Running the test in parallel

Let's go back a bit and see what we used in Chapter 4, *Understanding Desired Capabilities*: the Refactoring -2 section. Here's the code snippet we used:

```
@Before
public void startAppiumServer() throws IOException {

    int port = 4723;
    String nodeJS_Path = "C:/Program Files/NodeJS/node.exe";
    String appiumJS_Path = "C:/Program
    Files/Appium/node_modules/appium/bin/appium.js";

    String osName = System.getProperty("os.name");

    if (osName.contains("Mac")) {
        appiumService = AppiumDriverLocalService.buildService(new
        AppiumServiceBuilder()
                .usingDriverExecutable(new File("/usr/local/bin/node"))
                .withAppiumJS(new File("/usr/local/bin/appium"))
                .withIPAddress("0.0.0.0")
                .usingPort(port)
                .withArgument(GeneralServerFlag.SESSION_OVERRIDE)
                .withLogFile(new File("build/appium.log")));
    } else if (osName.contains("Windows")) {
        appiumService = AppiumDriverLocalService.buildService(new
        AppiumServiceBuilder()
                .usingDriverExecutable(new File(nodeJS_Path))
                .withAppiumJS(new File(appiumJS_Path))
                .withIPAddress("0.0.0.0")
                .usingPort(port)
                .withArgument(GeneralServerFlag.SESSION_OVERRIDE)
                .withLogFile(new File("build/appium.log")));
    }
    appiumService.start();
}
```

We discussed the ROBOT_ADDRESS capability, but didn't use it then. This capability holds the key to have Appium tests run in parallel.

We can follow these steps to implement test parallelization for Appium:

1. Create a method to start Appium service by passing port and udid as parameters.

2. Once we parameterize the preceding, we can actually start as many Appium servers as we have devices connected. The following code takes `port` and `udid` as parameters, starts the Appium service, and ties it to a particular `port` and `udid`:

```
appiumService = AppiumDriverLocalService.buildService(new
AppiumServiceBuilder()
.usingDriverExecutable(new File("/usr/local/bin/node"))
.withAppiumJS(new File("/usr/local/bin/appium"))
.withIPAddress("127.0.0.1")
.usingPort(port)
.withArgument(GeneralServerFlag.ROBOT_ADDRESS, udid as String)
.withArgument(AndroidServerFlag.BOOTSTRAP_PORT_NUMBER,
(port + 2) as String)
.withArgument(SESSION_OVERRIDE)
.withLogFile(new File("build/${udid}.log")));

appiumService.start();
```

3. Create a method to read the output of the `adb devices` command:
   - Iterate the preceding method to start the Appium service for each `udid` (Android device connected)
   - Use the following method to read the output of the preceding command:

```
public List<String> attachedDevicesAndEmulators() {
    List<String> devices = new ArrayList<>();
    String line;
    StringBuilder log = new StringBuilder();
    Process process;
    Runtime rt = Runtime.getRuntime();
    try {
        process = rt.exec(new String[]
        {"adb", "devices", "-l"});
        BufferedReader stdInput = new BufferedReader(new
        InputStreamReader(
                process.getInputStream()));
        BufferedReader stdError = new BufferedReader(new
        InputStreamReader(
                process.getErrorStream()));

        while ((line = stdInput.readLine()) != null) {
            log.append(line);
            log.append(System.getProperty
            ("line.separator"));
```

```
        }
        while ((line = stdError.readLine()) != null) {
            log.append(line);
            log.append(System.getProperty
            ("line.separator"));
        }
    } catch (Exception e) {
        e.printStackTrace();
    }
    Scanner scan = new Scanner(String.valueOf(log));
    while (scan.hasNextLine()) {
        String oneLine = scan.nextLine();
        if (oneLine.contains("model")) {
            devices.add(oneLine.split("device")[0].trim());
        }
    }
    return devices;
}
```

4. Create a method to build the desired capability based on the device UDID as the parameter
5. Create a properties file to save the mapping of tags and devices to pick at runtime
6. Create a method in Gradle to read from the properties file and run the test

For the preceding steps, you need to implement your own code.

Network conditioning

Mostly, we test a mobile app in a perfect condition of best and fast network; however, in reality the devices might be moving and the network may be fluctuating between Edge connections (2G), 3G, or even LTE. Sometimes, the automation test has to run at a lower data speed or even test some offline functionality.

Appium exposes the `driver.setConnection()` method, which can help in setting the network condition between WiFi, airplane, data, or none. Any of the following statements can be used, based on which data connectivity you want to set up:

```
driver.setConnection(Connection.AIRPLANE);
driver.setConnection(Connection.WIFI);
driver.setConnection(Connection.DATA);
driver.setConnection(Connection.NONE);
driver.setConnection(Connection.ALL );
```

The connection is an `enum` that defines these bit masks:

Connection Type	Bit Mask
NONE	0
AIRPLANE	1
WIFI	2
DATA	4
ALL	6

Once the value is set, it persists for the life of the driver instance, so we must reset it back to the data connectivity we want for the test suite.

On macOS, one can install the **Network Link Conditioner** app to simulate the various network conditions. It can be downloaded as part of the Hardware IO tools package (for more information visit: `https://developer.apple.com/download/more/?q=Hardware%20I O%20Tools`). The following screenshot shows what the app looks like. This helps simulate the network speed on the simulator. One thing to keep in mind is that it impacts the hosting device network speed as well, so we have to be careful while using it:

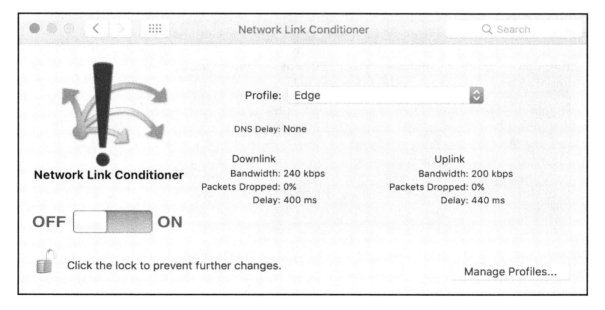

Profile lets you select between different network speeds, such as low latency and 3G:

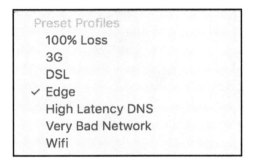

However, on a real iOS device, it's already built in and can be accessed by navigating to **Settings** > **Developer** > **Network link Conditioner**.

Summary

This chapter completes our journey learning mobile test automation with Appium. It took us on a tour where we understood the importance of mobile app testing and automation. We learned about the mobile testing ecosystem, how to set up a machine, and how to install the respective software and tools. We learned how to use the Appium app, find locators, and author tests. We also learned how to automate gestures and how to introduce synchronization in tests. We saw how to run these tests on devices and emulators, including setting up Genymotion emulators. We also discussed how to set up Jenkins and have tests automated when the source code is checked in Github.

Lastly, in this chapter, we learned some Appium tricks for switching between WebView and Native, taking screenshots, and embedding them in the report. We explored how to record the test execution device screen and also how to vary the quality of the recording. We learned how to interact with other apps and traverse back to the app under test. We learned the approach for parallel test execution and how to implement it. We also learned how to simulate the various network conditions to simulate 2G, 3G, or LTE conditions on the device while running the functional test.

With this knowledge, we are good to go out, set up our own automation framework from scratch, and drive it to solve our testing needs. I wish good luck and happy learning to you all!

12
Appium Desktop App

In this chapter, we will take a detailed look at the new Appium app that is built on Electron. We will look at how to use the app and different options it allows us to configure:

- Installing the Appium app
- Starting a simple server
- Starting the server with advanced options
- Appium endpoints

Installing the new Appium app

Appium has recently released a new open source GUI app for Mac/Windows/Ubuntu users. It's an app with a new UI and doesn't require node or NPM to be installed. It's built using electron (for more information visit: `https://electron.atom.io/`) and comes bundled with node runtime. It can be downloaded and installed from the specified location (for more information visit: `https://github.com/appium/appium-desktop/releases/tag/v1.0.0`). Based on your machine OS, you can choose to download the respective installer file and install the Appium app. The latest released version is **1.6.4**.

Here's how the new Appium app looks when you launch it after install:

At the first glance, it allows you to do the following:

- Start a simple server (version 1.6.4) with the default configuration
- Explore some additional settings under the **Advanced** tab, and then start the server or save it as presets

Let's take a detailed look at starting a simple server.

Starting a simple server

To start an Appium server, we only need the host and port info. The new app allows you to update the host and port information and then start the server. It also indicates the server version, which is `1.6.4` at the time of writing this book.

When you click on the **Start Server v1.6.4** button, it opens the console log that shows the status of the server:

```
[Appium] Welcome to Appium v1.6.4
[Appium] Appium REST http interface listener started on 0.0.0.0:4723
```

The following is the new interface when the server is running. It shows you the Appium server runtime log as well as other options:

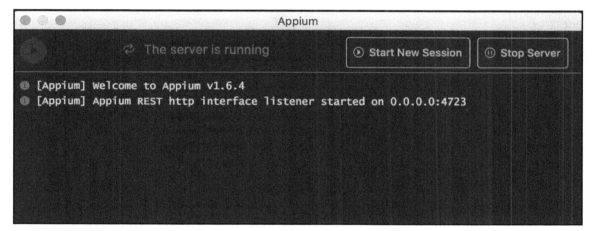

The app gives you two options:

- Start New Session
- Stop Server

Start New Session

Clicking on **Start New Session** launches a new screen (as shown in the following screenshot), which allows you to launch a new Appium session with the specified Desired Capabilities. By default, the new session will be launched against the default running server; alternatively, we can choose to use the other endpoints, such as **Custom Server**, **SauceLabs**, and **TestObject**. We will discuss that later:

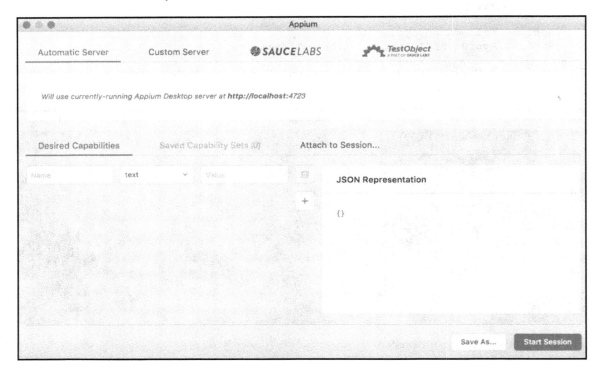

Attach to an existing session

It allows you to attach to an existing session by providing just the **session-id** (as shown in the following screenshot). This comes in handy when you already have an Appium session, and you are in the middle of a running test. Attaching to an existing Appium session is possible because the inspector is just an Appium client:

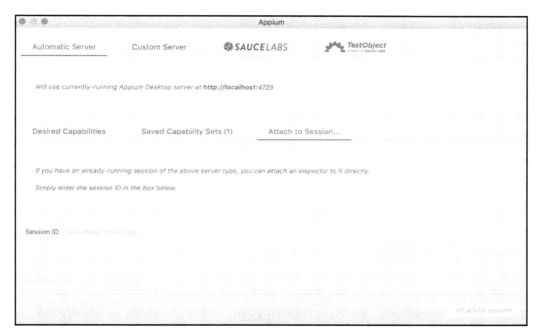

Desired Capabilities

Let's do the exercise of launching a new session. We need four mandatory Desired Capabilities to launch a new session when we are working with a pre-existing app or three mandatory Desired Capabilities when we want to deploy the app on emulator.

When we want to launch a new session for an installed app (Quikr, in our case) on emulator/device, use the mentioned Desired Capabilities:

- **platformName:** Android
- **deviceName**: Nexus
- **appPackage**: com.quikr
- **appActivity**: com.quikr.old.SplashActivity

Here's how the screen will look after setting the values. On the right-hand side, you can see the JSON being created when we add new **Desired Capabilities**:

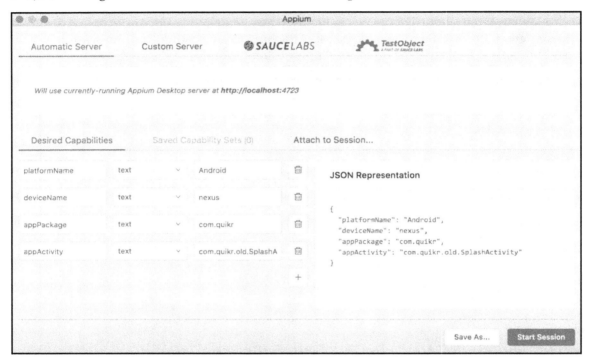

Clicking on **Start Session** will launch a new Appium inspector screen, as illustrated. Clicking on **Save As...** will allow you to save the config as preset values.

Appium Inspector

Once you click on **Start Session**, it launches the Appium Inspector, which is fairly simple to use; one needs to click on the element on the left-hand side of the screen and the right pane **Selected Element** will load to show the app source and details of the selected element. The right pane is categorized by **Find By** and the rest of the attributes of that element, such as index, text, and class:

It allows you to perform operations such as **Tap** and **Send Keys**:

It also allows you to navigate **Back** (which simulates the action on the device) and **Refresh** the UI based on the device's current state. Clicking on **Quit** closes the Appium inspector session.

Starting the server with advanced options

Appium app allows you to start the server with the **advanced options**. Clicking on **Advanced** on the launch screen opens a new configuration section in the app, which allows you to select the **General Server** arguments and the **iOS/Android** specific arguments:

It allows you to enter the following details:

- **General**:
  - **LogFile Path**: This is the location where we want to store the Appium log file.
  - **Log Level**: The default value is debug; other allowed values are info, info:debug, info:info, info:warn, info:error, warn, warn:debug, warn:info, warn:warn, warn:error, error, error:debug, error:info, error:warn, error:error, debug, debug:debug, debug:info, debug:warn, and debug:error.
  - **Override Temp Path**: This is the absolute path to the directory Appium can use to manage temporary files.
  - **Node Config File Path**: This is the configuration JSON file to register Appium.
  - **Local Timezone**: This is to use the local timezone for timestamps.
  - **Allow Session Override**: This enables session override.
  - **Log Timestamps**: They show timestamps in console output.
  - **Suppress Log Colour**: Do not use colors in console output.
  - **Strict Caps mode**: This causes sessions to fail if desired caps are sent, and it does not recognize it as valid for the selected device.

- **iOS**
  - **WebDriverAgent Port**: Local port used for communication with WebDriverAgent.
  - **executeAsync Callback Host**: Callback IP Address (default: the same as address).
  - **executeAsync Callback Port**: Callback port (default: the same as port).

- **Android**
  - **Bootstrap Port**: Port to use on device to talk to Appium.
  - **Selendroid Port**: Local port used for communication with Selendroid.
  - **Chromedriver Port**: Port upon which ChromeDriver will run. If not passed, Android driver will pick a random available port.
  - **Chromedriver Binary Path**: ChromeDriver executable full path.

Appium app allows you to the save the config by clicking on the **Save As Preset...** option.

Appium Endpoints

Appium app also allows you to launch a session against a non-local Appium server. There are built-in integrations with **SauceLabs** and **TestObject**, apart from running your server on a custom host.

- **Custom Server**: This allows you to launch an Inspector session against an Appium server running on another machine in your network. It allows you to provide the host address and the port:

- **Sauce Labs**: This allows you to leverage your Sauce Labs (for more information visit `https://saucelabs.com/`) account to start an Appium session in the cloud:

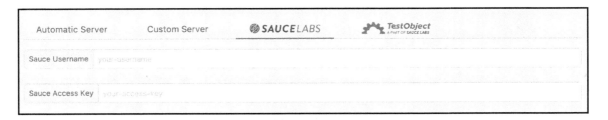

- **TestObject**: This allows you to leverage the cloud of real devices of TestObject (for more information visit `https://testobject.com/`):

Summary

In this chapter, we learned about the new Appium app and how to install it. We learned how to run a simple server and also learned how to start a new session using the Desired Capabilities and by attaching to an existing session. We saw an example to launch the existing Quikr app on the emulator and save the configurations as a preset.

We also learned to use the Appium inspector and the options it gives us, such as tap and send keys. We also looked at different options given in the Appium inspector, such as **Refresh**, **Quit**, and **Navigate back**. We explored how to use the advanced options to create an Appium session and different server arguments that Appium gives us to configure under the heading **General** and device-specific for **iOS and Android**.

We also learned about the integration with other endpoints, such as custom server, SauceLabs, and TestObject.

This chapter covers the new Appium app, which is still to be widely adopted and used.

Appendix

Introduction to Cucumber

In **Behavior Driven Development** (**BDD**), the prime focus is on writing acceptance tests that describe the behavior of the application or system. Acceptance tests are written from a customer point of view and hence bring in the outside-in approach to the understanding and testing of the application. The emphasis is on making the test cases readable by everyone on the team so that any stakeholder can give feedback on the application's behavior.

Eric Evans, in his book Domain Driven Design (`http://www.amazon.in/Domain-Driven-De` `sign-Tackling-Complexity-Software/dp/0321125215`), talks about the need for one language to bridge the gap between the domain experts and programmers on the team. Cucumber helps enforce the ubiquitous language within the team, which can be understood by anyone on the team. Cucumber tests are written in a language that can be understood by anyone in the team, and it's implementation tests the application. This way, Cucumber helps a team express the behavior of the application in a language that is executable, and at the same time, understandable by stakeholders.

Cucumber clearly makes it easy, given simplicity with which it can be authored and comprehended by anyone in the team. An example is as follows:

```
Feature: Car Search

   Scenario: Search for a used Honda City car in Bangalore city

      Given I launch the app
      When I choose "Bangalore" as my city
      And I search for "Honda City" under Used Cars
      Then I should see the first car search result with "Honda"
```

Now with this example, anyone would be able to comprehend what the behavior under test is. Also, it is very easy for others to ascertain whether we are testing the right scenario. The amazing aspect of Cucumber is that this feature is executable; it can be run and provides feedback.

Cucumber solves the problem of documentation and serves as a source of living specification of the software. Most of the time, the documentation resides in a system such as an excel sheet or some test case management system. The challenges of that approach are the maintenance and diligence required to keep it updated. The advantage with Cucumber is that it will always be updated, otherwise the test will fail. It never becomes outdated because of the constant maintenance and feedback it gives.

Cucumber also serves as a source of truth, being in one place that gives complete insight into the application's behavior. It takes away the pain as well as the time of maintaining multiple documents. It also helps in avoiding people having their own version of truth and understanding of the application.

How does Cucumber work?

Cucumber is a command-line tool that basically executes the feature file which contains business scenarios facing the application. Feature files follow a specific syntax that is called Gherkin. Gherkin is a **Domain Specific Language** (DSL) that allows us to describe a business scenario. It's a line-oriented language that uses spaces or tabs to define structure apart from the keyword.

There are two basic conventions with Gherkin:

- A file can contain the description of a single feature
- Files have the `.feature` extension

The stages of writing a scenario will be as follows:

1. Create a feature file.
2. Describe a scenario.
3. Write the steps to accomplish that scenario.

All these steps of writing a scenario are business facing, while the implementation is purely technical. Let's see a better representation of the Cucumber stack (pic courtesy: *The Cucumber Book*):

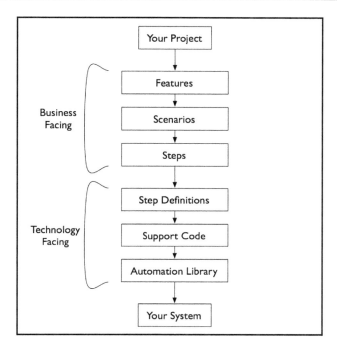

The technology facing component can be implemented in different languages such as Ruby, Java, .NET (using SpecFlow), and JavaScript. Here's a representation that makes it more clear:

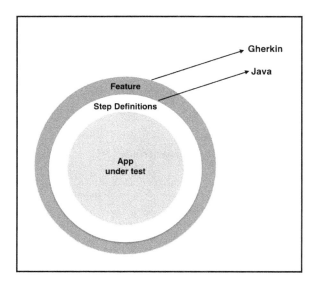

Let's take a quick deep dive into Gherkin and understand it in a bit more detail. The beauty of Gherkin lies in its simplicity to write a feature file. Feature files can be authored in any text editor tool available. It follows syntaxes such as **YAML Ain't Markup Language (YAML)**. A Gherkin file uses a `.feature` extension and can be created in any text editor. It starts with a `Feature` keyword and is written in plain English using other keywords. Let's take a look at the different keywords Gherkin has:

- Feature
- Background
- Examples
- Scenario
- Given
- When
- Then
- And
- Scenario Outline

Feature

Feature is the first keyword to be used in a Gherkin file. Each Gherkin file can have only one feature. The typical syntax is this:

```
Feature: This is feature name
This is feature description and
it can be multi-line till the Gherkin parser
encounters the next Keyword
```

So, the text following the Feature keyword is the feature name that expresses the business module under test; some examples of feature names are Login, Search, and User Registration. Feature description can be expressive and can detail what is supposed to be accomplished by that feature.

Gherkin parser treats the entire text under feature description till it encounters another Gherkin keyword beginning on a new line.

Scenario

Scenario is another Gherkin keyword that helps express the business scenario under test. It captures the high-level intent of the scenario. The typical syntax is as shown:

```
Scenario: Scenario name
```

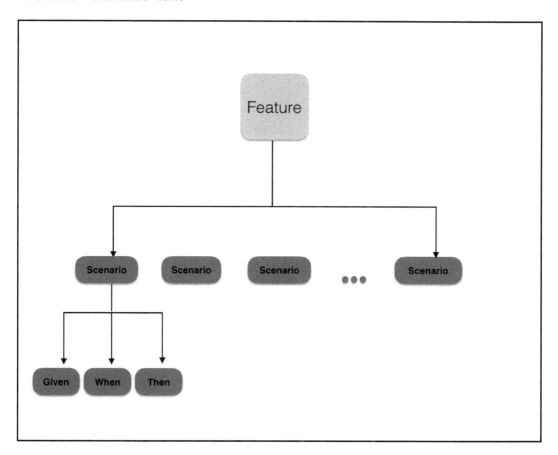

A feature can be broken down into multiple scenarios, and these scenarios constitute business use cases together. If you add up all the scenarios' behavior, it should be equivalent to the feature behavior itself.

So, a scenario basically contains the steps run on the system under test and gives feedback. For a scenario to pass in Cucumber, all steps under it should pass. Each scenario can have multiple steps describing the behavior. There is no rule for the number of steps within a scenario; however, care should be taken to keep the readability intact.

Gherkin gives us keywords to help express these steps; they are `Given`, `When`, `Then`. Any testing scenario is generally categorized into the following:

- Getting the system to a desired state
- Performing the steps to test
- Verifying

The mentioned steps are typically mapped to `Given`, where we get the system in a desired state, `When`, where we perform the actual testing steps (this can be a bunch of lines) and lastly, `Then`, where we do the verification of the desired state of the application. Let's look at the feature file we wrote earlier:

```
Given I launch the app
When I choose "Bangalore" as my city
And I search for "Honda City" under Used Cars
Then I should see the first car search result with "Honda"
```

So, `Given` sets the application state that is about launching the desired application; `When` is telling the application to move to a particular state by choosing a city and searching for specified cars in our case, and `Then` is about verifying that the first car result is the desired one.

Gherkin allows you to replace the entire set of Given, When, and Then in a little less verbose way by replacing it with `*`. So, the earlier statement can be expressed as follows:

```
* I launch iOS app
* I choose to enter "22" and "33"
* I tap on Compute Sum
* I should see the result "55"
```

Now that we have read about feature files, scenarios, and steps, let's take a look at the result states Cucumber gives. Cucumber has multiple states for the results: `Undefined`, `Pending`, `Passed`, and `Failed`.

Undefined steps: When Cucumber doesn't find the step definition that matches a step, it marks the step as undefined and throws the undefined step exception when we try to run it. For the preceding steps, it will throw the shown exception:

```
Undefined step: I launch iOS app
```

Pending steps: Cucumber isn't able to figure out whether a step is defined or not. It starts looking at the step definition and then figures out the state of the step, that is, whether it is defined or not. Generally, when a new step is created, this is the template:

```
@Given("^I launch iOS app$")
  public void iLaunchIOSApp() throws Throwable {
      // Write code here that turns the phrase above into concrete
      actions
      throw new PendingException();
  }
```

So, when the Cucumber runner encounters the `throw new PendingException()` statement, it throws up the pending steps exception.

Passed: If a code block executes successfully without throwing any exception, Cucumber marks that step as passed.

Failed: If a code block throws some exception, Cucumber marks that step as failed and skips the remaining steps (if any). The standard reason for exception is, generally, system not behaving as expected, which is a bug in the app or bug in the step definition code itself. The assertion failures also mark the step to be failed, thereby failing the scenario.

Let's look at another important Gherkin keyword--Background.

Background

Generally, while testing, we might have a bunch of scenarios that need a set of common steps. For example, any test steps that are after the log-in screen will require log in to be a common step. In that case, we can move log in to a section called `Background` in a feature file, thereby telling Cucumber to run it before each and every scenario in that file. Consider a feature file (testing the used car search scenario), as shown:

```
Feature: Used Car search feature

  Scenario: Search for a used Honda City car in Bangalore city

    When I launch Quikr app
    And I choose "Bangalore" as my city
    And I search for "Honda City" under Used Cars
    Then I should see the first car search result with "Honda"

  Scenario: Search for a used Honda City car in Bangalore city

    When I launch Quikr app
    And I choose "Bangalore" as my city
```

```
And I search for "Honda City" under Used Cars
And I select the budget to be 5L
Then I should see the first car search result with price less than
5L
```

Now, if we look at the preceding two scenarios, the first three steps are common for both the scenarios and are getting repeated. Instead, we can move some of these common steps to the background just below where Feature is mentioned under the Background keyword. Let's make the preceding changes and see the readability of the feature file:

```
Feature: Used Car search feature

Background:
    When I launch Quikr app
    And I choose "Bangalore" as my city


Scenario: Search for a used Honda City car in Bangalore city
    When I search for "Honda City" under Used Cars
    Then I should see the first car search result with "Honda"


Scenario: Search for a used Honda City car in Bangalore city
    When I select the budget to be 5L
    Then I should see the first car search result with price less than
    5L
```

So, we have added a Background section that takes care of setting the state of the application for both the scenarios and, in this case, it will perform the following steps:

- Launch the application under tests
- Choose Bangalore as the city for any further action

So, the purpose of the preceding two tests doesn't change; during runtime, Cucumber actually executes these background steps before each scenario. Part of the rule is that we can only have one Background per feature file and, secondly, it has to appear before the Scenario keyword or the Scenario Outline keyword.

Let's look at Scenario Outline.

Scenario Outline

In testing, we generally have scenarios where we have multiple combinations of input and different outputs for the same set of steps, such as the log-in combination and some other business calculation. `Scenario Outline` helps express these scenarios in a much better way by letting us express the scenario once and giving us an option to provide multiple sets of data in the `Examples` section. Let's take a look at the given example:

```
Feature: Log in


    Scenario: Log in - right email/password input
        Given I launch the app
        When I get the user sign in screen
        And I enter "valid@email.com" and "valid password"
        Then I should see a message "Log in success"


    Scenario: Log in - wrong email input
        Given I launch the app
        When I get the user sign in screen
        And I enter "wrong@email.com" and "valid password"
        Then I should see a message "In-valid email provided"

    Scenario: Log in - wrong password input
        Given I launch the app
        When I get the user sign in screen
        And I enter "valid@email.com" and "wrong password"
        Then I should see a message "Wrong password"
```

In the preceding scenario, we have the same set of steps repeating for different data combinations that are the essence of the test cases. We can express the same scenario in a much better way with less repetitiveness. Refer to the following usage of `Scenario Outline` to achieve this:

```
Scenario Outline: Log in combinations
    Given I launch the app
    When I get the user sign in screen
    And I enter <email> and <password>
    Then I should see a message <message>
    Examples:
        | email           | password       | message                 |
        | valid@email.com | valid password | Log in success          |
        | wrong@email.com | valid password | In-valid email provided |
        | valid@email.com | wrong password | Wrong password entered  |
```

What happens behind the scene is that Cucumber converts each example row as one scenario and executes it. So basically, the `<email>` is nothing but a place holder that is substituted by the real values during execution. In a feature file, we can have many `Scenario Outline` and `Examples` sections. If we create a `Scenario Outline` and don't include a following `Example` section, it will throw an error.

Hooks in Cucumber

Cucumber has a very interesting feature of hooks that helps us execute a block of code before or/and after each scenario. It can be defined anywhere in the step definitions using the `Before` and `After` methods. Most of the xUnit tools support a concept of the setup and tear down method, which is represented by `Before` and `After` here.

By default, these hooks are global in nature, and they run for every scenario. Here, an interesting concept to understand is that the step definitions are global in nature; there is no way to reduce the scope of step definitions to certain scenarios.

A sample of the `Before` hook is as shown:

```
@Before
public void setUp() throws IOException {
    System.out.println("This is a set up method and will be called
    before the scenario");
}
```

A sample of the `After` hook is as follows:

```
@After
public void tearDown() throws IOException {
    System.out.println("This is a tear down method and will be called
    after the scenario");
}
```

Running Cucumber

Cucumber allows you to run feature files in a couple of ways:

- CLI Runner
- JUnit Runner
- Third-Party Runner (IntelliJ IDEA)

CLI Runner

CLI Runner stands for Command-Line Interface Runner, which is an executable class and can be invoked from Gradle or Ant. While using Cucumber-jvm on the command line, we can use this command:

```
java -cp <classpath> cucumber.api.cli.Main \
    --glue com.example.steps \
    --plugin pretty path/to/feature/files
```

JUnit Runner

If we are using the **JUnit** framework to run Cucumber, we need to create a single empty class, as shown:

```
package steps;

import cucumber.api.junit.Cucumber;
import org.junit.runner.RunWith;

@RunWith(Cucumber.class)
@CucumberOptions(plugin = {"pretty", "html:target/cucumber"})
public class RunCukesTest {
}
```

With this, we can run the tests in the same way as we run the typical **JUnit** tests.

Third-Party Runner (Via IntelliJ)

IntelliJ enables you to run Cucumber features via the `cucumber.cli.main` class. Navigate to **IntelliJ** > **Run** > **Edit Configurations**; we can configure it as shown in the following screenshot:

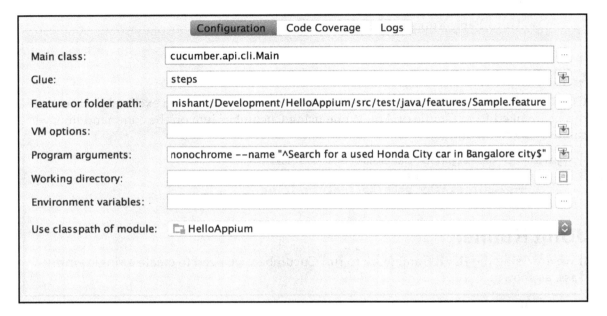

Let's look at some of these important items in detail:

- **Main**: This is the `main()` method, the main class name is `cucumber.cli.main`.
- **Glue**: This is the package name where the step definitions are contained.
- **Feature or folder path**: This is the directory name where the feature file is contained. You can also specify a specific feature here.
- **VM Options**: This is the string value to be passed to the VM for launching the app. The string contains options such as `mx` and `verbose`. If we specify a class path here as part of VM Options, it will override the class path of the module.
- **Program Arguments**: This is the list of arguments to be passed to the program in the same format as that of the command line.

If you are using **Eclipse**, it also provides similar options to run Cucumber-ivm test.

Finding an app's package name and launch activity

In `Chapter 5`, we entered `Package name` and `Launch Activity` to launch Appium Inspector for an app already installed on the emulator. Let's learn how to find this information from an app.

We can follow two approaches to get the same result. The first approach requires you to have **Play Store** and the app under test (**Quikr** in our case) installed on your mobile.

Using the ManifestViewer app

Follow the given steps to find out the package information:

- Launch the Emulator/Device
- Launch **Google Play Store**
- Search for an app **ManifestViewer** in Play Store and install it:

- Once installed, launch the **ManifestViewer** app

- Under the **Application** sections and scroll down to the Quikr app:

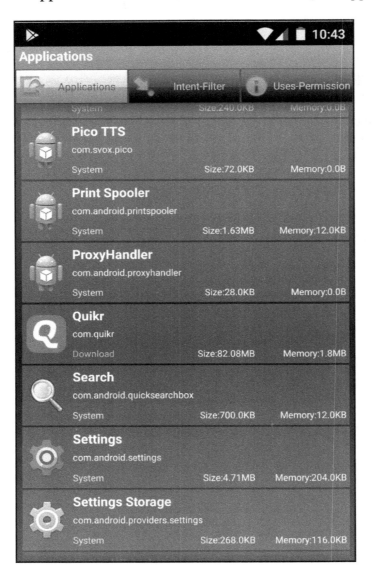

- Once done, tap on the **Quikr** app, and it will show you these options:

- Click on the Look the `AndroidManifest.xml` option
- This will load the manifest properties, as illustrated:

- This file will have details like **package** and **activity**

Using the Appium GUI app

The second approach to know the package and activity details is to use the Appium GUI app. When we use Appium to install the app on the emulator/device, it also loads the **Package** details and the **Launch Activity**. So, the steps to be followed are as listed:

1. Launch the **Emulator/Device.**
2. Launch the **Appium GUI** app.
3. Select the **App path** parameter and browse to the **APK package.**
4. Click on **Launch.**
5. This will start the Appium server; now click on the **Inspector** icon.
6. Click on **Stop** (to stop the Appium server).
7. Click on the Android icon in the Appium GUI app.
8. Select the **Package** checkbox and click on the dropdown; it will show the value from the last APK file installed.
9. Select the **Launch Activity** checkbox and click on the dropdown; it will show all the values from the last APK file installed.

Refer to the following screenshot:

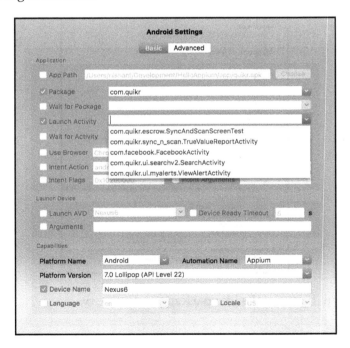

Installing Google Play services in the Genymotion emulator

Genymotion is one of the fastest Android emulators available for use. One drawback of using Genymotion is that it comes without the Google Play Store and Google apps. This means that some of the apps for testing that require the Google Play services framework may not work on the emulator.

However, the good news is that we can install Google Play services by following these steps:

- Start the Genymotion emulator
- Based on the Android version configured for the emulator, we need to download the flash-able Google Play services `gapps-lp-yyyymmdd-signed.zip` file from

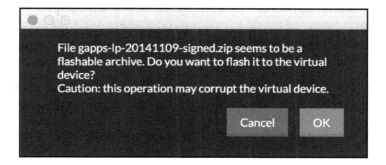

This installs Google Play services on the Genymotion emulator.

- Click on **OK** and reboot the emulator
- Once the device is rebooted, you will notice that the Google Apps will start showing in the emulator
- To install Google Play Store on the device, we need to download the Play Store installer `com.android.vending-Major.Minor.Hotfix` APK file and install it on the emulator

- Once done, launch the Play Store app and log in to the Play Store account (if you have one)
- This will update all the necessary Google apps (some of the apps might intermittently crash or stop working, but this will only occur until the apps are updated)
- Restart the emulator once the apps are updated, and it will work smoothly

Summary

In this Appendix, we covered different topics for a deeper insight into Cucumber. We learned about how Cucumber works and the importance of BDD, and we gained a deep insight into Gherkin and the different keywords Gherkin exposes. We also learned what hooks are and how to use them. We learned the different ways of running Cucumber tests.

We also learned how to look up an Android package name and find out different activities for an app. This is needed when we want to launch the Appium session on a pre-existing app on an Android device. We also learned how we can find the package name and Launch Activity from Appium itself.

We learnt that the Genymotion emulator doesn't come with Google Play services installed. We learned how to flash the device with the Google apps installer file and to install Google Play services on the Genymotion emulator.

Index